Measurement Unit Conversions
A complete workbook with lessons and problems

By Maria Miller

Copyright 2017 Maria Miller.
ISBN 978-1545173466

EDITION 4/2017

All rights reserved. No part of this workbook may be reproduced or transmitted in any form or by any means, electronic or mechanical, or by any information storage and retrieval system, without permission in writing from the author.

Copying permission: Permission IS granted for the teacher to reproduce this material to be used with students, not commercial resale, by virtue of the purchase of this workbook. In other words, the teacher MAY make copies of the pages to be used with students. Permission is given to make electronic copies of the material for back-up purposes only.

Contents

Preface	5
Introduction	7
Helpful Resources on the Internet	7
Decimals in Measuring Units and More	11
Rounding and Estimating	15
The Metric System	17
Convert Metric Measuring Units	20
Converting Between Customary Units of Measurement	23
Convert Customary Measuring Units	27
Convert Between Customary and Metric	31
Using Ratios to Convert Measuring Units	33
Maps	37
Significant Digits	43
Review	45
Answers	47
Appendix: Common Core Alignment	59

Preface

Hello! I am Maria Miller, the author of this math book. I love math, and I also love teaching. I hope that I can help you to love math also!

I was born in Finland, where I also grew up and received all of my education, including a Master's degree in mathematics. After I left Finland, I started tutoring some home-schooled children in mathematics. That was what sparked me to start writing math books in 2002, and I have kept on going ever since.

In my spare time, I enjoy swimming, bicycling, playing the piano, reading, and helping out with Inspire4.com website. You can learn more about me and about my other books at the website MathMammoth.com.

This book, along with all of my books, focuses on the conceptual side of math... also called the "why" of math. It is a part of a series of workbooks that covers all math concepts and topics for grades 1-7. Each book contains both instruction and exercises, so is actually better termed *worktext* (a textbook and workbook combined).

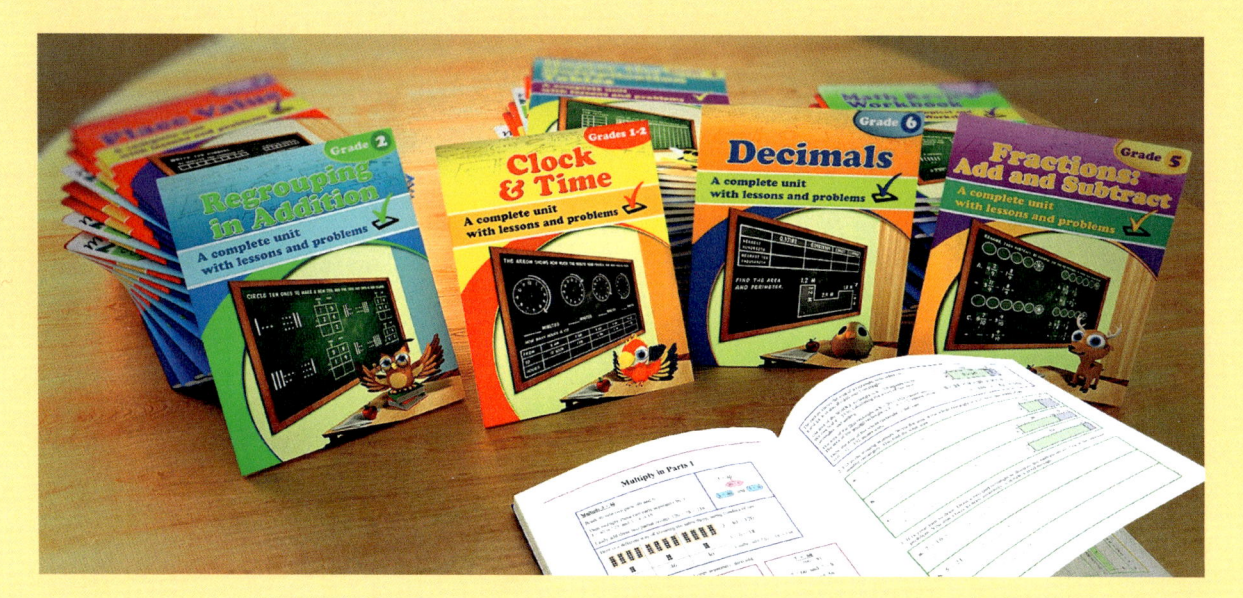

My lower level books (approximately grades 1-5) explain a lot of mental math strategies, which help build number sense — proven in studies to predict a student's further success in algebra.

All of the books employ visual models and exercises based on visual models, which, again, help you comprehend the "why" of math. The "how" of math, or procedures and algorithms, are not forgotten either. In these books, you will find plenty of varying exercises which will help you look at the ideas of math from several different angles.

I hope you will enjoy learning math with me!

Introduction

The workbook *Measurement Unit Conversions* contains lessons and exercises suitable for grades 5-7.

First, we study how the basic concept of decimal numbers can help us convert measuring units. For example, since 0.01 means one-hundredth, then 0.01 m means one-hundredth of a meter -- which is the definition of a *centi*meter. Then we round and estimate quantities given in various measurement units, and find the error of estimation.

After that, we focus on the metric system and conversions between the metric units of measurement. I have tried to emphasize sensible and intuitive methods for converting measuring units within the metric system, instead of relying on mechanical formulas.

Next, we practice conversions between units in the customary system, using the basic conversion factors and multiplication and division. Then from there we advance to conversions between customary and metric measuring units.

Later, students learn how rates can be used to convert measurement units. This method is in addition to the methods for converting measurement units that were explained earlier in the unit. It does not mean that students should "change over" and forget what they learned earlier—it is simply a different method for doing the conversions. Some students may choose one method over another; some may be able to master all of the methods. Most will probably choose one method they prefer for doing these conversions.

Next, the lesson on maps gives a different real-world context for measurement units and conversions between them, since calculating the real distances from map distances or vice versa involves both using a scale ratio and conversion of the measurement unit used.

Lastly, the lesson *Significant Digits* deals with the concept of the accuracy of a measurement and how it limits the accuracy of the solution. Significant digits is not a standard topic for middle school, yet the concept in it is quite important, especially in science. You can consider this topic as optional or reserve it for advanced students.

I wish you success with your math teaching!

Maria Miller

Helpful Resources on the Internet

Use these free online resources to supplement the "bookwork" as you see fit.

UNITS OF MEASUREMENT

Conversion Quizzes - ThatQuiz.org
A customizable online quiz about conversions between measuring units. The options include both metric and customary systems and six different difficulty levels.
http://www.thatquiz.org/tq-n/science/metric-system/

Horrendous Soup Game
Make a recipe for the most disgusting soup you can imagine in this fun game that practices conversion between metric units of measurement.
http://mrnussbaum.com/soup

Metric System Conversions Quiz
Practice converting between different units of measurement in the metric system with this 10-question online quiz.
https://www.thatquiz.org/tq-n/?-j17v-l4-p0

Matthew Metric Bubble Gum Game
Practice metric and standard units of measurement while filling the customers' bubble gum orders.
http://mrnussbaum.com/matthewmetric/

Word Problems Involving Measurement Conversions
Solve word problems that involve converting between metric measures of distance, volume, and mass, as well as measures of time.
https://www.khanacademy.org/math/on-sixth-grade-math/on-measurement/on-unit-conversion/e/converting-measurements-word-problems

Common Conversion Factors Practice
Practice memorizing the common conversion factors in this interactive exercise.
http://www.sporcle.com/games/12121/measurement-conversion

Convert Mixed Customary Units
Practice converting customary units of measurement in this interactive online exercise.
http://www.mathgames.com/skill/5.10-convert-mixed-customary-units

Convert Customary Units
Fill in the tables to convert between US customary measures of distance, volume, and mass.
https://www.khanacademy.org/math/cc-fifth-grade-math/cc-5th-measurement-topic/cc-5th-unit-conversion/e/converting-units--us-customary-

Customary Unit Conversion Printable Worksheets
Use these randomly generated worksheets for extra practice. Refresh the page (F5) to get a different worksheet.
http://www.homeschoolmath.net/worksheets/measuring-customary.php#grade5

Converting Units - Word Problems
Solve word problems that involve converting between US customary measures of distance, volume, and mass in this interactive exercise from Khan Academy.
https://www.khanacademy.org/math/cc-fifth-grade-math/cc-5th-measurement-topic/cc-5th-unit-word-problems/e/converting-units-word-problems--us-customary-

USING RATIOS TO CONVERT MEASUREMENT UNITS

Unit Conversion Tool
Use this interactive tool for all types of unit conversion. Includes a slider that you can adjust to see various conversions.
http://www.mathsisfun.com/unit-conversion-tool.php

Converting Units with Dimensional Analysis
This page includes explanations, videos, and exercises to practice converting units.
https://www.texasgateway.org/resource/converting-between-measurement-systems

Ocean Math Worksheet
This ocean-themed worksheet contains a set of measurement conversion problems to solve.
http://www.mathgoodies.com/worksheets/ocean_wks.html

SCALE DRAWINGS AND MAPS

Ratio and Scale
An online unit about scale models, scale factors, and maps with interactive exercises and animations.
http://www.absorblearning.com/mathematics/demo/units/KCA024.html

Scale Drawings and Maps Quiz
Answer questions about scales on maps and scale drawings in these five self-check word problems.
http://www.math6.org/ratios/8.6_quiz.htm

Maps
A tutorial with worked out examples and interactive exercises about how to calculate distances on the map or in real life based on the map's scale.
http://www.cimt.org.uk/projects/mepres/book7/bk7i19/bk7_19i3.htm

Short Quiz on Maps
Practice map-related concepts in this multiple-choice quiz.
http://www.proprofs.com/quiz-school/story.php?title=map-scales

SIGNIFICANT DIGITS

Sig Fig Rules
Drag Sig J. Fig to cover each significant digit in the given number.
http://www.sigfig.dreamhosters.com/

Practice on Significant Figures
A multiple-choice quiz that also reminds you of the rules for significant digits.
http://www.chemistrywithmsdana.org/wp-content/uploads/2012/07/SigFig.html

Significant digits quiz
A 10-question multiple-choice quiz about significant digits.
http://www.quia.com/quiz/114241.html?AP_rand=1260486279

Dimensional Analysis Quiz
Use the conversions given in the table to help you answer the questions in this multiple-choice quiz.
http://ths.sps.lane.edu/chemweb/unit1/problems/dimensionalanalysis/

Decimals in Measuring Units and More

> Since one meter is 100 cm, one-tenth of a meter is 10 cm: **0.1 m = 10 cm**.
> Then, 0.7 meters is 7 times that, or 70 cm.
>
> Similarly, one-hundredth of a meter is 1 cm: **0.01 m = 1 cm**.

1. Write as centimeters or meters.

a. 0.8 m = _____ cm	**b.** 56 cm = _____ m	**c.** 0.31 m = _____ cm
1.3 m = _____ cm	3 cm = _____ m	_____ m = 460 cm
8.27 m = _____ cm	382 cm = _____ m	16.08 m = _____ cm

1 km is 1,000 m. Therefore:	Similarly:	**Example 1.** How many meters is 0.032 km?
• 0.1 km is 100 m; • 0.01 km is 10 m; • 0.001 km is 1 m.	• 0.4 km is 400 m; • 0.05 km is 50 m; • 0.472 km is 472 m.	This is 32/1000 of a kilometer. Each 1/1000 of a km is one meter, so 32/1000 km is equal to 32 meters.

2. Write as meters or kilometers.

a. _____ km = 500 m	**b.** 2,400 m = _____ km	**c.** 2.001 km = _____ m
0.7 km = _____ m	12,680 m = _____ km	0.319 km = _____ m
4.5 km = _____ m	540 m = _____ km	0.04 km = _____ m

3. Mark is 1.88 m tall, and 16 cm taller than Jake.
   How tall is Jake?

4. Jack ran around a 550-meter track four times.
   What distance did he run in kilometers?

5. From Ann's home to the swimming pool is 2.4 km, and from the pool to the library is 350 m.
   Ann rode her bicycle to the pool, then to the library, then back to the pool and back home.
   What distance did she travel in total?

| 1 L is 1,000 ml. Therefore:

• 0.1 L is 100 ml;
• 0.01 L is 10 ml;
• 0.001 L is 1 ml. | Similarly:

• 0.3 L is 300 ml;
• 0.07 L is 70 ml;
• 0.923 L is 923 ml. | **Example 2.** Write 2.23 liters as milliliters.

Two liters is 2,000 milliliters. Then, 0.23 liters is 23/100 L, and each 1/100 L is 10 ml. So, 23/100 L is 23 × 10 ml = 230 ml. All in all, 2.23 L = 2,230 ml. |

6. Write as liters or milliliters.

a. 0.7 L = _____ ml	**b.** 3,900 ml = _____ L	**c.** 0.009 L = _____ ml
12.6 L = _____ ml	2,080 ml = _____ L	1.35 L = _____ ml
0.06 L = _____ ml	212 ml = _____ L	1.585 L = _____ ml

7. A shampoo bottle contains 0.47 L of shampoo. Another bottle contains 520 ml of shampoo. Find the difference in the amount of shampoo in the two bottles.

| **1 kg is 1,000 g.** After practicing with kilometers, meters, liters, and milliliters, you should be able to convert between kilograms and grams easily, since they work in a similar manner. | • 0.1 kg is _____ g

• 0.01 kg is _____ g

• 0.001 kg is _____ g |

8. Write as kilograms or grams.

a. 0.3 kg = _____ g	**b.** 20 g = _____ kg	**c.** 1.1 kg = _____ g
2.6 kg = _____ g	800 g = _____ kg	0.152 kg = _____ g
0.05 kg = _____ g	6,030 g = _____ kg	2.093 kg = _____ g

9. A large apple weighs 0.18 kg. Joe ate 3/4 of it. How many grams of the apple did Joe eat?

10. Find the total weight, in kilograms, of 30 jars of honey weighing 450 grams each.

Example 3. One mile is 5,280 ft. How many feet is 0.6 miles?

The phrase "six tenths of a mile" can be written as a *decimal multiplication*, replacing "a mile" with 5,280 ft. After multiplying, we get the answer of 3,168 feet.

0.6 of a mile
↓ ↓ ↓
0.6 × 5,280 ft →

```
    1 4
  5 2 8 0
×     0.6
---------
3 1 6 8.0
```

Example 4. Convert 0.95 of a mile into yards.

We use the same idea. The phrase "0.95 of a mile" can be written as a multiplication, and one mile is 1,760 yards. We get 0.95 × 1,760 yd = 1,672 yd.

0.95 of a mile
↓ ↓ ↓
0.95 × 1,760 yd →

```
      6 5
      3 3
    1 7 6 0
×     0.9 5
-----------
    8 8 0 0
+ 1 5 8 4 0 0
-------------
  1 6 7 2.0 0
```

11. Convert to feet or to yards. Don't use a calculator. *Round* your answers to whole feet or yards.

    a. 0.2 mi = _____ ft

    b. 1.35 mi = _____ ft

    c. 2.7 mi = _____ yd

    d. 0.72 mi = _____ yd

12. A road maintenance crew completed 0.7 mi of road on Monday, 0.65 mi on Tuesday, and 0.5 mi on each of the remaining three weekdays. Find how much road they completed in the week, both in miles and in feet.

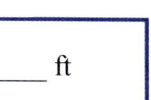

13. Write as feet or yards. Now use a calculator. Round your answers to whole feet or yards.

a. 0.65 mi = _____ ft	b. 0.9 mi = _____ yd	c. 5.428 mi = _____ ft
1.34 mi = _____ ft	5.413 mi = _____ yd	2.75 mi = _____ yd

14. Mount McKinley is 20,320 feet tall. The International Space Station flies 211.3 miles above the Earth. How many Mount McKinleys would you need to stack on top of each other in order to reach the altitude of the International Space Station?

One-tenth or 0.1 of a million dollars is $100,000.	0.01 (one-hundredth) of $1,000,000 is $10,000.
So, 0.6 of a million dollars would be $600,000. And, 5.6 million dollars is $5,600,000.	0.37 of a million dollars is 37 times that, or $370,000.

15. Write as dollars. Here "M" means "million dollars."

a. $0.7 M = $ _____	**b.** $0.04 M = $ _____	**c.** $10.9 M = $ _____
$2.5 M = $ _____	$0.39 M = $ _____	$2.78 M = $ _____

16. The city of Luisberg planned to spend $2.85 M on public education in the year 2012, but ended up spending about $350,000 less. How much did they spend on public education?

17. The tables give examples of how the world record for long jump has progressed over the years. The last line gives the current record.

    Which has advanced more in centimeters from 1960 until the current record, the women's or the men's record?

    How much more?

Men's World Record Progression

Mark	Athlete	Date
8.21 m	Ralph Boston (USA)	August 12, 1960
8.35 m	Igor Ter-Ovanesyan (URS)	October 19, 1967
8.90 m	Bob Beamon (USA)	October 18, 1968
8.95 m	Mike Powell (USA)	August 30, 1991

Women's World Record Progression

Mark	Athlete	Date
6.40 m	Hildrun Claus (GDR)	August 7, 1960
6.70 m	Tatyana Shchelkanova (URS)	July 4, 1964
7.09 m	Vilma Bardauskiené (URS)	August 29, 1978
7.43 m	Anişoara Cuşmir (ROU)	June 4, 1983
7.52 m	Galina Chistyakova (URS)	June 11, 1988

18. You are placing chairs that are 55 cm wide in a row in a large dining room. The room is 9 m wide. If you leave 90 cm walking space at both ends of your row, how many chairs can you fit in the row?

Rounding and Estimating

Example 1. Jack needs 187 cm of board for his woodworking project. About how many meters is that?

To round 187 cm to the nearest meter, we need to remember that 1 m is 100 cm, so rounding to the nearest meter means rounding to the <u>nearest hundred centimeters</u>: 187 cm ≈ 200 cm = 2 m. Jack needs about 2 m of board. The **rounding error**, or the difference between these two numbers, is 13 cm or 0.13 m.

What if we round 187 cm to the nearest *tenth* of a meter? First, write 187 cm in meters. It is 1.87 m. Then round 1.87 m ≈ 1.9 m. The rounding error is now only 3 cm or 0.03 m.

1. Round these lengths to the nearest meter. Find the rounding error.

a. 8.19 m ≈ _____ m; rounding error _____	**b.** 362 cm ≈ _____ m; rounding error _____	**c.** 417 cm ≈ _____ m; rounding error _____
d. 1 m 54 cm ≈ _____ m; rounding error _____	**e.** 14.208 m ≈ _____ m; rounding error _____	**f.** 8 m 9 cm ≈ _____ m; rounding error _____

2. Round these lengths to the nearest kilometer. Find the rounding error.

a. 602 m ≈ _____ km; rounding error _____	**b.** 10.189 km ≈ _____ km; rounding error _____	**c.** 8.057 km ≈ _____ km; rounding error _____
d. 2643 m ≈ _____ km; rounding error _____	**e.** 6 km 55 m ≈ _____ km; rounding error _____	**f.** 3288 m ≈ _____ km; rounding error _____

3. Calculate. Round the final answer to the nearest tenth of a meter.

a. 1.5 m − 67 cm	**b.** 6.08 m + 45 cm + 1.2 m	**c.** 1 m 8 cm + 2.55 m

4. Calculate and round the final answer to the nearest ten meters.

a. 2.1 km − 293 m	**b.** 6.07 km + 452 m	**c.** 2 km 75 m + 3.8 km

5. Use rounded numbers to estimate the answer.
   Then find the exact answer using a calculator.

| a. A book is 0.78 inches thick. How many of those fit into a box 4 inches high? Estimation: Exact answer: | b. How many 0.024 kg papers can you mail with a 0.400 kg weight limit? Estimation: Exact answer: |

6. If one paperclip is 2.2 cm long, *estimate* the total length of 18 paperclips.

7. Jack knows that the width of his fathom, or fully outstretched arms from fingertip to fingertip, is 146 cm. Jack measures the width of a room using his fathom, and measures it as about 2 ½ of his fathoms. Estimate the width of the room in meters.

8. A jogging track is 1.425 km long. How far does Eric jog if he goes around it three times? Give your answer to the nearest 100 meters.

9. Jim had $50 when he went shopping. He bought ten yards of material for $2.46 per yard. How much money does Jim have left?

   a. Choose an expression that matches the problem.

   b. Fill in the missing dollar amount in the model.

   c. Estimate how much money Jim has left now.

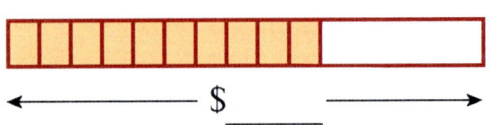

   $50 − ($2.46 ÷ 10)

   $50 − ($2.46 + 10)

   $50 − (10 × $2.46)

   d. Find the exact amount that Jim has left now.

The Metric System

The basic unit of length in the metric system is the **meter**. All the other units of length are formed by adding a prefix to the word "meter." For example, in "centimeter," the prefix is *centi*, which signifies 1/100. So, a centi-meter is 1/100 of a meter.

We can convert quantities that use units with prefixes back to the basic unit by translating the prefix.

Example 1. Convert 26 cm into meters. Since *centi* signifies a hundredth, 26 centimeters is 26 hundredths of a meter, or 0.26 m.

Example 2. Since *hecto* signifies 100, 7 hectometers is 7 hundred meters, or 700 m.

Units of Length in the Metric System

kilometer	km	1,000 meters
hectometer	hm	100 meters
dekameter	dam	10 meters
meter	m	the basic unit
decimeter	dm	1/10 of a meter
centimeter	cm	1/100 of a meter
millimeter	mm	1/1000 of a meter

(each step is ×10)

1. Write the amounts in the basic unit, meters, by "translating" the prefixes.

a. 2 cm = *2/100 m* = *0.02 m*	b. 3 dam = _____ m	c. 6 mm = _____ m
6 dm = _____ m = _____ m	9 km = _____ m	20 cm = _____ m
8 mm = _____ m = _____ m	2 hm = _____ m	8 dm = _____ m

Those same prefixes are used with *all* metric units, including the liter, the gram, the volt, and so on. Again, we can "translate" the prefix into the number it signifies if we want to convert a quantity to the basic unit.

Example 3. Eight deciliters (8 dl) means 8 tenths of a liter, or 0.8 liters, because a *deci* signifies one tenth.

Example 4. Five hectograms (5 hg) means 500 grams since *hecto* signifies a hundred.

Some of the units formed with the prefixes, such as dekagrams or hectometers, are not widely used. The most common prefixes are milli-, centi-, kilo-, and mega- (*mega* means 1,000,000).

Prefix	meaning
kilo-	1,000
hecto-	100
deka-	10
-	(the basic unit)
deci-	1/10
centi-	1/100
milli-	1/1000

2. Write the amounts in liters or grams by "translating" the prefixes.

a. 2 ml = *2/1000 L* = *0.002 L*	b. 7 dl = _____ L	c. 3 dag = _____ g
6 cl = _____ L = _____ L	6 mg = _____ g	8 kg = _____ g
8 dg = _____ g = _____ g	8 dl = _____ L	2 hl = _____ L

Metric units work just like place value. Let's write the quantity 6,734 cm into the metric unit chart. We place a decimal point after the "ones place," which currently is at centimeters.

km	hm	dam	m	dm	cm	mm
			6	7	3	4.

To convert 6,734 cm to any other unit in the chart, simply move the decimal point right after that unit. Don't move the numbers! For example, to change 6,734 cm to meters, we move the decimal point right after the meters place, and it becomes 67.34 m.

km	hm	dam	m	dm	cm	mm
			6	7.	3	4

To convert this measurement to hectometers, we move the decimal point just after hectometers. We also need to place a zero for hectometers. We get 0.6734 hm.

km	hm	dam	m	dm	cm	mm
	0.	6	7	3	4	

Example 5. Convert 46.7 dm to kilometers.

km	hm	dam	m	dm	cm	mm
			4	6.	7	

→

km	hm	dam	m	dm	cm	mm
0.	0	0	4	6	7	

Write 46.7 in the chart so that "6", which is in the ones place, is placed in the decimeters place.

Move the decimal point right after the kilometers place. Add necessary zeros. Answer: 0.00467 km.

3. Write the measurements in the metric unit charts.

   a. 75.4 m

   b. 843 mm

   c. 4.6 km

   d. 35.49 dam

4. Convert the measurements to the given units, using the charts above.

	to m	to dm	to cm	to mm
a. 75.4 m				
b. 843 mm				

5. Convert the measurements to the given units, using the charts above.

	to hm	to dam	to m	to dm
a. 4.6 km				
b. 35.49 dam				

6. In the following measurements, what measuring unit does the digit "3" correspond to?
   You may use the chart to help.

   a. 5.392 kg   b. 14.3 dg   c. 15,389 mg

   | kg | hg | dag | g | dg | cg | mg |

7. Write the measurement given on the right as

   a. deciliters   b. liters

   c. dekaliters   d. hectoliters

   ```
   4  5  0  0
   ↑  ↑  ↑  ↑
   hl dal L  dl
   ```

8. Convert the lengths in the table into the given measuring units.

	a. 5,000 mm	b. 380 cm	c. 6.5 dm
meters			
decimeters			
centimeters			
millimeters			

9. If you are supposed to take 2 cl of medicine daily out of a 200-ml bottle, how many days will the bottle last?

   | kl | hl | dal | l | dl | cl | ml |

10. Hannah is 120 cm tall, and Erica is 1.05 m tall. At what height would the tops of their heads be, if the girls stood on stools with the heights of:

    a. 3.1 dm

    b. 550 mm

    c. 45 cm

11. A paper clip weighs 14 dg. They are sold in boxes of 200.

    a. Calculate the weight of the box, in grams.

    | kg | hg | dag | g | dg | cg | mg |

    b. If someone wanted 1 kg of paperclips, how many boxes would they need to buy?

Convert Metric Measuring Units

The metric system has one basic unit for each thing we might measure: For length, the unit is the **meter**. For weight, it is the **gram**. And for volume, it is the **liter**.

All of the other units for measuring length, weight, or volume are *derived* from the basic units using *prefixes*. The prefixes tell us what multiple of the basic unit the *derived unit* is.

For example, centiliter is 1/100 part of a liter (*centi* means 1/100).

Prefix	Abbreviated	Meaning
kilo-	k	1,000
hecto-	h	100
deka-	da	10
-	-	(the basic unit)
deci-	d	1/10
centi-	c	1/100
milli-	m	1/1000

Unit	Abbr	Meaning
kilometer	km	1,000 meters
hectometer	hm	100 meters
dekameter	dam	10 meters
meter	m	(the basic unit)
decimeter	dm	1/10 meter
centimeter	cm	1/100 meter
millimeter	mm	1/1000 meter

Unit	Abbr	Meaning
kilogram	kg	1,000 grams
hectogram	hg	100 grams
dekagram	dag	10 grams
gram	g	(the basic unit)
decigram	dg	1/10 gram
centigram	cg	1/100 gram
milligram	mg	1/1000 gram

Unit	Abbr	Meaning
kiloliter	kl	1,000 liters
hectoliter	hl	100 liters
dekaliter	dal	10 liters
liter	L	(the basic unit)
deciliter	dl	1/10 liter
centiliter	cl	1/100 liter
milliliter	ml	1/1000 liter

1. Write these amounts using the basic units (meters, grams, or liters) by "translating" the prefixes. Use both fractions and decimals, like this: 3 cm = 3/100 m = 0.03 m (since "centi" means "hundredth part").

   a. 3 cm = *3/100 m* = *0.03 m*

   5 mm = _____ m = _____ m

   7 dl = _____ L = _____ L

   b. 2 cg = _____ g = _____ g

   6 ml = _____ L = _____ L

   1 dg = _____ g = _____ g

2. Write the amounts in basic units (meters, grams, or liters) by "translating" the prefixes.

   a. 3 kl = _____ L

   8 dag = _____ g

   6 hm = _____ m

   b. 2 dam = _____ m

   9 hl = _____ L

   7 kg = _____ g

   c. 70 km = _____ m

   5 hg = _____ g

   8 dal = _____ L

3. Write the amounts with derived units (units with prefixes) and a single-digit number.

   a. 3,000 g = _3_ _kg_

   800 L = _8_ _____

   60 m = _6_ _____

   b. 0.01 m = _____ _____

   0.2 L = _____ _____

   0.005 g = _____ _____

   c. 0.04 L = _____ _____

   0.8 m = _____ _____

   0.007 L = _____ _____

4. Write using prefixed units.

   a. 0.04 meters = 4 cm
   b. 0.005 grams = 5 _____
   c. 0.037 meters = 37 _____
   d. 400 liters = 4 _____
   e. 0.6 meters = 6 _____
   f. 2,000 meters = 2 _____
   g. 0.206 liters = 206 _____
   h. 20 meters = 2 _____
   i. 0.9 grams = 9 _____

5. Change into the basic unit (either meter, liter, or gram). Think of the meaning of the prefix.

   a. 45 cm = *0.45 m*
   b. 65 mg =
   c. 2 dm =
   d. 81 km =
   e. 6 ml =
   f. 758 mg =
   g. 2 kl =
   h. 8 dl =
   i. 9 dag =

Example 1. Convert 2.5 cg to grams.

					2.	5
kg	hg	dag	g	dg	cg	mg

→

				0.	0	2	5
kg	hg	dag	g	dg	cg	mg

Write 2.5 in the chart so that "2", which is in the ones place, is placed in the centigrams place.

Move the decimal point just after the grams place. Add necessary zeros. Answer: 0.025 g.

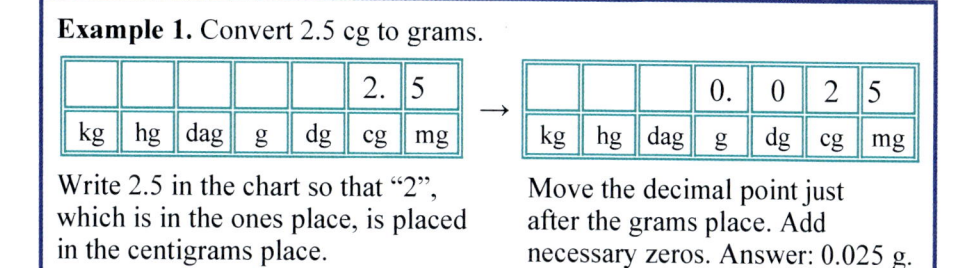

6. Write the measurements in the place value charts.

   a. 12.3 m   [km | hm | dam | m | dm | cm | mm]

   b. 78 mm   [km | hm | dam | m | dm | cm | mm]

   c. 56 cl   [kl | hl | dal | l | dl | cl | ml]

   d. 9.83 hg [kg | hg | dag | g | dg | cg | mg]

7. Convert the measurements to the given units, using the charts above.

	m	dm	cm	mm
a. 12.3 m	12.3			
b. 78 mm				78 mm
	L	dl	cl	ml
c. 56 cl				
	g	dg	cg	mg
d. 9.83 hg				

21

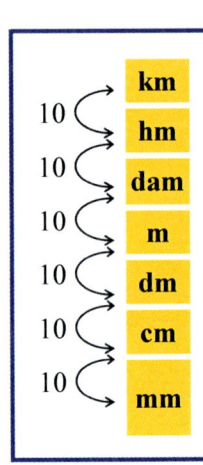

You can also convert measurements by thinking of how many steps apart the two units are in the chart and then multiplying or dividing by the corresponding power of ten.

Example 2. Convert 2.4 km into centimeters.

There are five steps from kilometers to centimeters. That means we would multiply 2.4 by 10, five times — or multiply 2.4 by 10^5.

2.4 × 100,000 = 240,000, so 2.4 km = 240,000 cm.

Example 3. Convert 2,900 cg into hectograms.

"Centi" and "hecto" are four steps apart, so we will divide by 10^4 = 1000.
2,900 ÷ 10,000 = 0.29, so 2,900 cg = 0.29 hg.

8. Convert the measurements. You can write the numbers in the place value charts or count the steps.

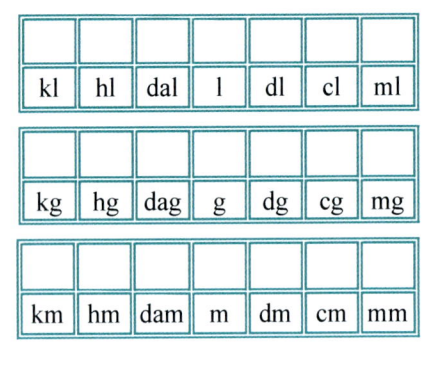

   a. 560 cl = _____ L

   b. 0.493 kg = _____ dag

   c. 24.5 hm = _____ cm

   d. 491 cm = _____ m

   e. 35,200 mg = _____ g

   f. 32 dal = _____ cl

   g. 0.483 km = _____ dm

   h. 0.0056 km = _____ cm

   i. 1.98 hl = _____ dl

   j. 9.5 dl = _____ L

9. Each measurement has a flub, either in the unit or in the decimal point. Correct them.

   a. The length of a pencil: 13 m

   b. The length of an eraser: 45 cm

   c. Circumference of Dad's waist: 9.2 m

   d. The height of a room: 0.24 m

   e. Jack's height: 1.70 mm

   f. Jenny's height: 1.34 cm

10. Find the total ...

    a. ... weight of books that weigh individually:
    1.2 kg, 1.04 kg, 520 g, and 128 g.

    b. ... volume of containers whose individual volumes are:
    1.4 L, 2.25 L, 550 ml, 240 ml, and 4 dl.

11. A dropper measures 4 ml. How many full droppers can you get from a 2-dl bottle?

12. Once a day, a nurse has to give a patient 3 mg of medicine for each kilogram of body weight. The patient weighs 70 kg. How many days will it take for the patient to take 2 g of medicine?

Converting Between Customary Units of Measurement

Units of weight

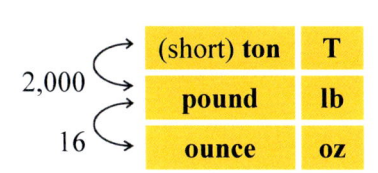

2,000 → (short) ton — T
→ pound — lb
16 → ounce — oz

Units of volume

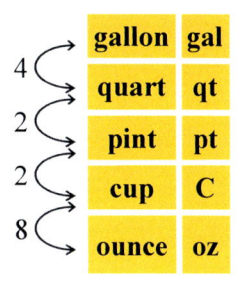

4 → gallon — gal
→ quart — qt
2 → pint — pt
2 → cup — C
8 → ounce — oz

Units of length

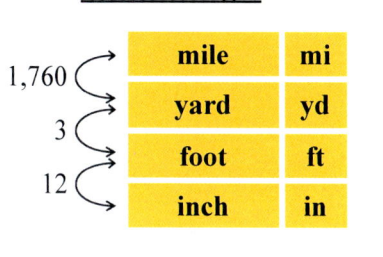

1,760 → mile — mi
→ yard — yd
3 → foot — ft
12 → inch — in

To convert from one neighboring unit to another, either **multiply** or **divide by the conversion factor**.

If you don't know which, THINK if the result needs to be a smaller or bigger number.

Example 1. Convert 53 ounces into cups.

The conversion factor we need is 8, because 8 ounces makes a cup (look at the chart). And, ounces are smaller units than cups, so 53 ounces as cups will make *fewer* cups (you need fewer cups since they are the bigger units). So, we need to divide by the factor 8:

53 ÷ 8 = 6 R5. The result means 53 ounces is 6 cups and 5 (leftover) ounces.

You can also think of it this way: since 8 ounces makes a cup, we need to figure how many cups or how many "8 ounce servings" there are in 53 ounces. How many 8s are in 53? That is solved by division.

1. Convert.

a. 6 ft = _____ in.	b. 25 in = _____ ft _____ in	c. 13 ft 7 in = _____ in
7 ft 5 in. = _____ in.	45 in = _____ ft _____ in	71 in = _____ ft _____ in

2. Convert.

a. 2 lb 8 oz = _____ oz	b. 8 lb = _____ oz	c. 43 oz = _____ lb _____ oz
45 oz = ____ lb _____ oz	56 oz = _____ lb _____ oz	90 oz = _____ lb _____ oz

3. Convert.

a. 3 C = _____ oz	b. 4 C = _____ pt	c. 7 gal = _____ qt
55 oz = _____ C _____ oz	3 pt = _____ C	45 qt = _____ gal _____ qt

23

Example 2. Convert 3 quarts into ounces.

We are going from bigger units (quarts) to smaller units (ounces), so there will be *more* of them. We need to multiply.

This time, quarts and ounces are not neighboring units in the chart. We need to **multiply 3 quarts by all of the factors** between them: by 2, by 2, and by 8. We get 3 qt = 3 × 2 × 2 × 8 oz = 96 oz.

	gallon	gal
4	quart	qt
2	pint	pt
2	cup	C
8	ounce	oz

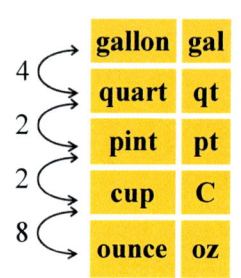

Example 3. Convert 742 inches into yards.

We can do the conversion in two steps: first into feet and then into yards.

From inches to feet, the conversion factor is 12. Since feet are bigger units than inches, we need fewer of them, so we will divide. See the long division at the right. The result is that 742 ÷ 12 = 61 R 10. It means that 742 in = 61 ft 10 in.

Then we convert 61 ft into yards by dividing by 3 (the 10 inches will not count in this conversion since they don't make even one yard):

61 ÷ 3 = 20 R1. So, 61 ft = 20 yd 1 ft.

Lastly we put it all together: all in all, 742 in = 20 yd 1 ft 10 in.

```
      0 6 1
12 )7 4 2
     7 2
     2 2
   - 1 2
     1 0
```

4. Convert. Use long division or multiplication.

   a. 11 yd = _____ in.

   b. 711 in. = _____ ft _____ in

   c. 982 in. = _____ yd _____ ft _____ in

   d. 254 oz. = _____ C _____ oz

   Now, convert the cup-amount of your answer above into quarts and cups.

   254 oz. = _____ qt _____ C _____ oz

   Lastly, convert the quart-amount into gallons and quarts.

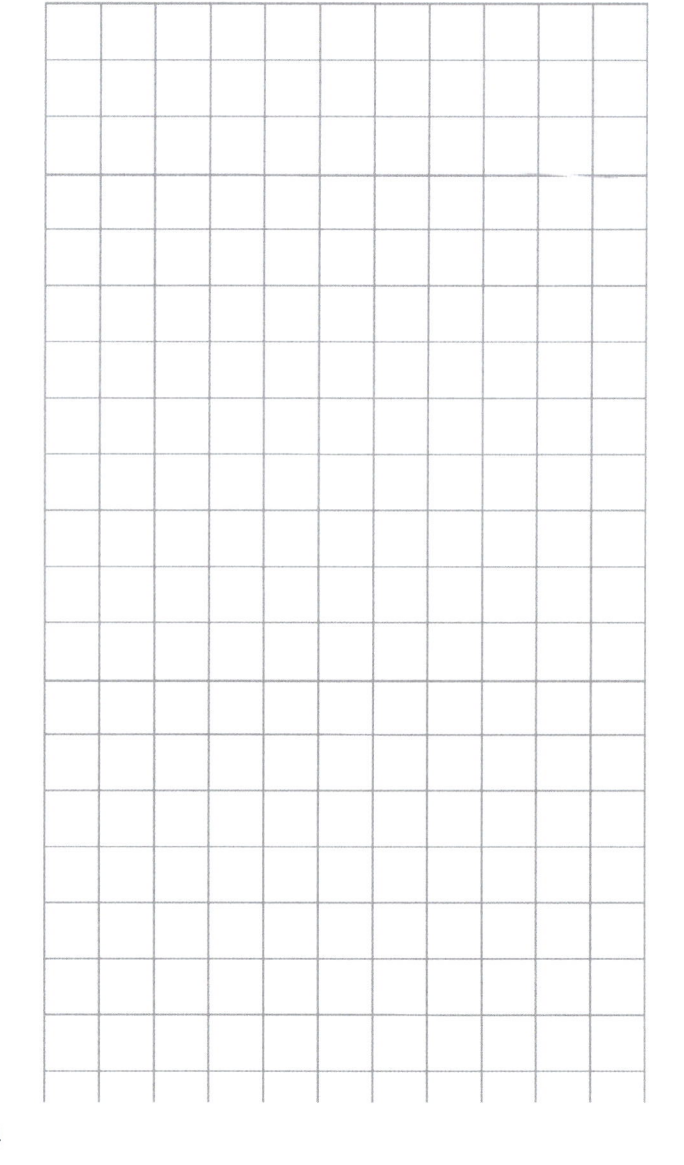

   254 oz. = _____ gal _____ qt _____ C _____ oz

Example 4. Convert 4.52 lb into ounces.

We are going from bigger units (pounds) to smaller units (ounces), so there will be lots more of them. We need to *multiply*.

Using a calculator, we get 4.52 × 16 = 72.32 oz.

2,000 →	(short) ton	T
16 →	pound	lb
	ounce	oz

Example 5. How many miles is 8,400 feet?

Since one mile is 5,280 feet, then 8,400 feet would be somewhere between 1 and 2 miles.
To find out exactly, use division, and round the answer: 8,400 ÷ 5,280 = 1.59090909... ≈ 1.59 miles.

5. Convert. Use a calculator. Round your answer to two decimal digits, if necessary.

a. 7.4 mi = _____ ft	b. 1,500 ft = _____ yd
16,000 ft = _____ mi	7,500 yd = _____ mi
c. 900 ft = _____ mi	d. 12.54 mi = _____ ft
2.56 mi = _____ yd	82,000 ft = _____ mi

1,760 → mile mi
3 → yard yd
12 → foot ft
  → inch in

1 mile = 5,280 feet

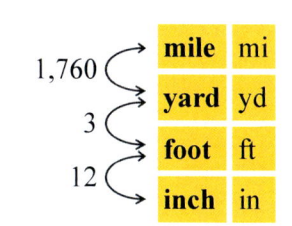

6. Convert. Use a calculator. Round your answer to two decimal digits, if necessary.

| a. 15.2 lb = _____ oz | b. 4.78 T = _____ lb | c. 78 oz = _____ lb |
| 655 oz = _____ lb | 7,550 lb = _____ T | 0.702 T = _____ lb |

7. Solve the riddle. Use the calculator for the problems that you cannot solve in your head.

F 0.6 mi = _____ ft   G 7 C = _____ oz   I 14,256 ft = _____ mi
A 5,632 yd = _____ mi   R 6,200 lb = _____ T   W 6 ft 7 in = _____ in
O 10 qt = _____ C   S 3 lb 5 oz = _____ oz L 732 in = _____ ft
H 2 lb 11 oz = _____ oz   E 5 ft 2 in = _____ in D 42 in = _____ ft
L 1.3 mi = _____ yd   O 40 oz = _____ lb   P 3 gal = _____ pt
                                                          A 0.75 mi = _____ ft

What did one potato chip say to the other?

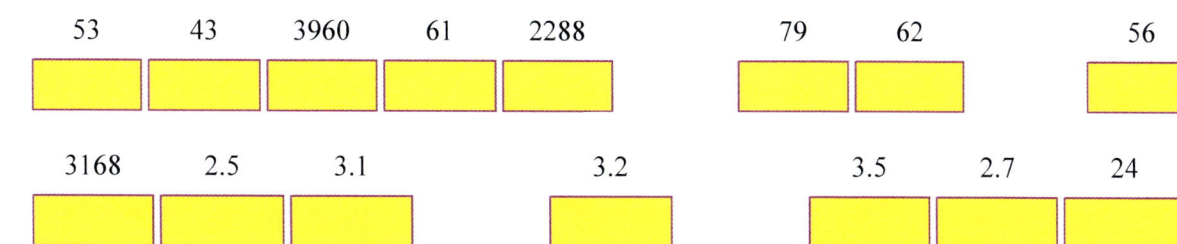

53   43   3960   61   2288   79   62   56   40
☐   ☐   ☐   ☐   ☐   ☐   ☐   ☐   ☐

3168   2.5   3.1   3.2   3.5   2.7   24
☐   ☐   ☐   ☐   ☐   ☐   ☐   ?

8. Solve.

   a. If you serve 1-cup servings of juice to 30 people,
   how many *whole* gallons of juice will you need?

   b. Mom was making applesauce in 2-gallon batches
   and bottling it in 1-quart jars. After 9 batches,
   how many jars of applesauce had she made?

   c. How many 8-inch pieces can you cut out of 9 3/4 ft of ribbon?

   d. A 4-ounce serving of coffee costs $1.20.
   What would a 5-ounce serving cost?

   e. A bottle of shampoo weighs 13 oz, and there are 20 of them in a box.
   The box itself weighs 8 oz. How much does the box with the bottles
   of shampoo weigh in total, in pounds and ounces?

   f. Mark drinks three 5-ounce servings of coffee a day.
   Find how much coffee he drinks in a month (30 days).
   Give your answer in bigger units, not in ounces.

   g. Erica lost 5 lb of weight over 4 weeks of time.
   How much weight did she lose daily, on average?

Convert Customary Measuring Units

Units of length

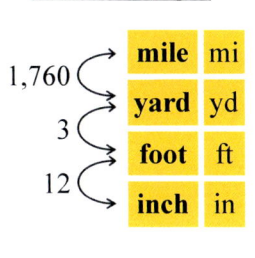

1,760 → mile | mi
3 → yard | yd
12 → foot | ft
→ inch | in

1 mile = 5,280 feet

Units of weight

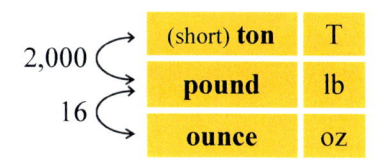

2,000 → (short) ton | T
16 → pound | lb
→ ounce | oz

Units of volume

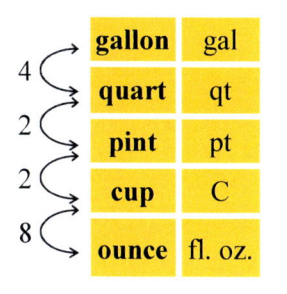

4 → gallon | gal
2 → quart | qt
2 → pint | pt
8 → cup | C
→ ounce | fl. oz.

When you convert between units, you either <u>multiply</u> or <u>divide</u> by the conversion factor. But which?

If the unit that you end with is *smaller* than the unit that you start with, then there should be *more* of them, and the number will get *bigger*. Use multiplication.

Conversely, if the unit that you end with is *bigger* than the unit that you started with, then there should be *fewer* of them, and the number will get *smaller*. Divide.

Example 1. Convert 11 ounces to pounds.

Ounces are *smaller* units than pounds, so there should be *more* of them. In fact, 11 ounces is less than 1 pound. Obviously, we have to divide: 11 ÷ 16 = 0.6875. We get 11 oz ≈ 0.69 lb.

Instead of a decimal, you could give this answer as a fraction very simply: 11 oz = $\frac{11}{16}$ lb.

Example 2. Convert 56,000 inches to miles.

Miles are a lot bigger than inches, so we expect to end up with fewer of them. In other words, we expect the number 56,000 to get *smaller*, so we will need to divide.

You can convert from inches to miles in two steps: first from inches to feet, then from feet to miles. The unit keeps getting bigger, so we keep dividing to get fewer of them.

56,000 ÷ 12 ÷ 5,280 = 0.88383838... So 56,000 inches ≈ 0.88 miles.

1. Which conversion is correct — the upper or the lower?

a. 2.46 gal = 2.46 × 4 qt = 9.84 qt 2.46 gal = $\frac{2.46}{4}$ qt = 0.615 qt	**b.** 11 oz = 11 × 16 lb = 176 lb 11 oz = $\frac{11}{16}$ lb = 0.6875 lb
c. 450 ft = 450 × 5,280 mi = 2,376,000 mi 450 ft = $\frac{450}{5,280}$ mi ≈ 0.085 mi	**d.** 12.6 ft = 12.6 × 12 in = 151.2 in 12.6 ft = $\frac{12.6}{12}$ in = 1.05 in

2. Convert to the given unit. Round your answers to two decimals, if needed.

a. 564 ft = _____ mi	**c.** 3,400 yd = _____ mi	**e.** 0.28 mi = _____ ft
b. 45,000 ft = _____ mi	**d.** 7.8 mi = _____ ft	**f.** 10.17 mi = _____ yd

You can use a calculator for all the problems in this lesson.

3. Convert to the given unit. Round your answers to two decimals, if needed.

a. 3 in = _____ ft	c. 14.7 ft = _____ in	e. 281 in = _____ ft
b. 21 in = _____ ft	d. 0.8 ft = _____ in	f. 7 1/3 ft = _____ in

4. Convert to the given unit. Round your answers to two decimals, if needed.

a. 5 oz = _____ lb	c. 3.6 lb = _____ oz	e. 127 oz = _____ lb
b. 35 oz = _____ lb	d. 0.391 lb = _____ oz	f. 6 3/4 lb = _____ oz

5. Convert to the given unit. Round your answers to two decimals, if needed.

a. 6.4 gal = _____ qt	d. 0.56 qt = _____ fl. oz.	g. 0.054 T = _____ lb
b. 78 fl. oz. = _____ qt	e. 560 qt = _____ gal	h. 1,200 lb = _____ T
c. 2.3 qt = _____ fl. oz.	f. 3.2 T = _____ lb	i. 6,750 lb = _____ T

Example 3. Convert 6 lb 15 oz into ounces.

Simply change the 6 lb into ounces first, then add the 15 ounces.

Example 4. Convert 372 ounces into pounds and ounces.

For the pounds, figure out how many 16-ounce increments there are in 372. That is done by dividing 372 ÷ 16. If you use long division, you will have a remainder, and the remainder tells you the individual ounces that are "left over." If you use a calculator, you will get a decimal number: 372 ÷ 16 = 23.25. The whole pounds are 23.

For the ounces, you can take the decimal part, 0.25, and figure out how many ounces 0.25 lb is. Another way is to calculate 23 × 16 = 368, and since that is 4 less than 372, there are four ounces.

In summary, 372 oz = 23 lb 4 oz.

6. Convert to the given unit. Round your answers to two decimals, if needed.

a. 2 ft 6 in = _____ in	c. 162 in = ____ ft _____ in	e. 254 in = ____ ft _____ in
b. 7 ft 11 in = _____ in	d. 79 in = ____ ft _____ in	f. 1,028 in = ____ ft _____ in

7. Convert to the given unit. Round your answers to two decimals, if needed.

a. 6 lb 9 oz = _____ oz	c. 86 oz = ____ lb _____ oz	e. 483 oz = ____ lb _____ oz
b. 11 lb 12 oz = _____ oz	d. 145 oz = ____ lb _____ oz	f. 591 oz = ____ lb _____ oz

Example 3. Convert 2.45 pounds to pounds and ounces.

This time, the 2 from 2.45 gives us the pounds. But the ounces? We need to convert the decimal part, 0.45 pounds, into ounces.

One pound is 16 ounces. Therefore, 0.45 pounds is 0.45 × 16 = 7.2 ounces. Most of the time, we give weights using whole pounds and whole ounces, so this would be rounded to 7 ounces. So, 2.45 lb is about 2 lb 7 oz.

8. Convert to the given unit. Round your answers to whole inches and whole ounces.

| a. 2.7 ft = ____ ft _____ in | c. 3.15 ft = ____ ft _____ in | e. 55.46 lb = ____ lb _____ oz |
| b. 10.2 ft = ____ ft _____ in | d. 7.8 lb = ____ lb _____ oz | f. 8.204 lb = ____ lb _____ oz |

With the worksheet maker at http://www.homeschoolmath.net/worksheets/measuring.php you can make more conversion problems between measuring units for extra practice.

9. You can add, subtract, and even multiply customary measuring units in columns.

a.
```
    5 lb  14 oz
+   7 lb  13 oz
```

b.
```
   34 ft   6 in
   62 ft   9 in
+  11 ft  11 in
```

c.
```
    6 qt  24 oz
    1 qt   7 oz
    4 qt  18 oz
+   2 qt  13 oz
```

d.
```
   60 ft   2 in
−  14 ft   8 in
```

e.
```
    2 lb   7 oz
×          5
```

f.
```
    5 h  34 min
−   2 h  45 min
```

10. Right now Jack is 4 feet 3 inches tall. He has been growing steadily at the rate of 2 3/8 inches per year for three years. How tall was he three years ago?

11. How many 21-inch wide chairs can you put in a row in a room that is 40 ft wide?

What if you wish to leave two 3-foot aisles?

12. Jack made 4 quarts of peppermint tea. He wants
    to serve it in small glasses in 6-ounce servings.
    How many glasses does he need?

13. You are packing math books that weigh 2 lb 3 oz each
    into a box that must not weigh more than 60 pounds.
    How many books can you put into the box?

14. Find the better deal.

    a. A 13-oz bottle of shampoo for $5.69 or
    a 1-quart bottle of shampoo for $13.99.

    b. 12 oz of potatoes for $0.35 or
    8 lb of potatoes for $4.10

15. A bottle of olive oil contains 25.5 oz and costs $4.84.
    What is the price per quart?

16. **a.** A gallon of ice cream is divided evenly among 17 people.
    How much does each person get (in ounces)?

    b. Your ice cream scoop holds 1.5 oz. How many scoops do
    you need to give each person so everyone gets an equal share?

17. You bought 10 pounds of strawberries and divided them evenly among
    seven people. How much did each person get (in pounds and ounces)?

Convert Between Customary and Metric

EASY ballpark figures:	Good to remember also:	Exact figures:	
1 m ≈ 1 yd	1 in ≈ 2.5 cm	1 inch = 2.54 cm	1 quart = 0.946 L
1 L ≈ 1 qt	1 mi ≈ 1.6 km	1 foot = 0.3048 m	1 ounce = 28.35 g
1 kg ≈ 2 lb	(4 laps on a 400-m track)	1 yard = 0.9144 m	1 lb = 0.454 kg
	1 oz ≈ 30 g	1 mile = 1.6093 km	1 kg = 2.2 lb

Example 1. Convert 17 inches to centimeters.

The table lists 1 inch as 2.54 cm. Therefore, 17 inches is simply 17 times that, or 17 × 2.54 cm = 43.18 cm.

Example 2. Convert 650 grams to ounces.

The table does *not* list 1 gram as so many ounces. If it did, we would multiply. Instead, it says that **1 ounce = 28.35 g**. So, each 28.35 grams makes one ounce. We *divide* to find how many times 28.35 grams fits into 650 grams—and that is the amount of ounces.

650 g ÷ 28.35 g/oz ≈ 22.93 oz or about 23 oz.

You can also see the need for dividing once you notice that ounces are bigger units than grams (1 ounce is about 28 grams). We need *fewer* of a bigger unit. Therefore, the number in 650 grams must get a lot smaller when it is converted into ounces.

In each conversion, you either multiply or divide by the conversion factor.

For example, the table gives 1 yard = 0.9144 m. If you have to convert yards into meters, multiply by 0.9144. To convert meters into yards, divide.

1. Circle the conversion on the right that is closest in size to the given measurement on the left.

a. 1 inch	1 cm 2.5 cm 5 cm	**b.** 1 foot	5 cm 10 cm 30 cm	**c.** 1 mile	1.5 km 2.5 km 3.5 km	**d.** 1 qt	1 L 100 ml 2.5 L
e. 2 kg	2 lb 4 lb 6 lb	**f.** 1 m	12 in 3 ft 3 yd	**g.** 1 cup	5 ml 30 ml 240 ml	**h.** 1 gal	4 L 6 L 8 L

2. Which is more? (Write <, =, or > between the measurements.)

a. 1 cm 1 in	**b.** 1 L 1 qt	**c.** 1 kg 1 lb	**d.** 1 g 1 oz
e. 4 in 20 cm	**f.** 5 kg 20 lb	**g.** 3 gal 2 L	**h.** 7 m 4 ft

3. Convert between the units. Use a calculator when needed. Round your answers to two decimals.

a.	b.	c.	d.
1 cm = _____ in	1 m = _____ yd	2 L = _____ qt	5 kg = _____ lb
25 cm = _____ in	5.4 m = _____ ft	4.6 L = _____ qt	0.568 kg = _____ lb

e.	f.	g.	h.
5 in = _____ cm	30 ft = _____ m	1 gal = _____ L	75 lb = _____ kg
10 in = _____ cm	22 ft = _____ m	3 1/2 qt = _____ L	8.5 lb = _____ kg

4. In the U.S., a common speed limit is 55 miles per hour.
   This corresponds most closely to a European speed limit of:
   (a) 70 km/h   (b) 80 km/h   (c) 90 km/h   (d) 100 km/h

5. Which is a better deal, a 24-ounce bottle of honey for $6.75
   or a 1-liter bottle of honey for $9.25?
   *Hint: For both bottles, find either the price per ounce or the price per liter.
   Note that here the term "ounces" refers to fluid ounces of volume,
   rather than ounces of weight.*

6. On the label of a food container, you can often find
   its capacity. A container's label reads 64 oz.

   a. Is it bigger than one that is 2.2 L?

   b. If so, how much larger? If not, how much smaller?

7. Angela weighs 56 kg, Theresa weighs 128 lb, Judy
   weighs 137 lb, and Elizabeth weighs 60 kg.
   List the girls in order from the lightest to the heaviest.

8. One marathon is 26.21875 miles.
   How long in kilometers is a half marathon?

Using Ratios to Convert Measuring Units

Consider the conversion factor **1 inch = 2.54 cm**. If we think of it as an equation and divide both sides by "1 inch," then we will get 1 on the left side, and a RATIO on the right side:

1 inch = 2.54 cm	This is the conversion factor, but we will think of it as an equation now.
$\dfrac{1 \text{ inch}}{1 \text{ inch}} = \dfrac{2.54 \text{ cm}}{1 \text{ inch}}$	Divide both sides by "1 inch." Yes, we do include the unit *inch* in this.
$1 = \dfrac{2.54 \text{ cm}}{1 \text{ inch}}$	We get a plain 1 on the left side (something divided by itself equals 1).

What we get on the right side is the ratio <u>2.54 cm per 1 inch</u> (or 2.54 cm to 1 inch), and that ratio equals 1.

We can also do this the other way around:

1 inch = 2.54 cm	This is the conversion factor, but we will think of it as an equation now.
$\dfrac{1 \text{ inch}}{2.54 \text{ cm}} = \dfrac{2.54 \text{ cm}}{2.54 \text{ cm}}$	Divide both sides by "2.54 cm". Yes, we do include the unit *cm* in this.
$\dfrac{1 \text{ inch}}{2.54 \text{ cm}} = 1$	We get a plain 1 on the right side (something divided by itself equals 1).

What we get on the left side is the ratio <u>1 inch per 2.54 cm</u> (or 1 inch to 2.54 cm), and that ratio equals 1.

In fact, we can transform *any* conversion factor between measuring units into a ratio that is equal to 1.

1 qt = 0.946 L	0.946 L = 1 qt	1 mi = 1.6093 km	1 lb = 0.454 kg
↓	↓	↓	↓
$\dfrac{1 \text{ qt}}{0.946 \text{ L}} = 1$	$\dfrac{0.946 \text{ L}}{1 \text{ qt}} = 1$	$\dfrac{1 \text{ mi}}{1.6093 \text{ km}} = 1$	$\dfrac{1 \text{ lb}}{0.454 \text{ kg}} = 1$

1. Think of the conversion factors as equations, and transform each one into a new equation of the form "1 = a ratio" or "a ratio = 1."

1 ft = 0.3048 m	1 ounce = 28.35 g	1 mi = 1,760 yd	1 m = 1.0936 yd
↓	↓	↓	↓

We can use these ratios that equal one in **converting measuring units**.

How does that happen? Study the following example carefully. Mathematically speaking, we multiply the quantity we want to convert by 1. Multiplying it by 1 does not change its value. Then, we replace that 1 with one of the ratios of measuring units that equal 1. Next, we cross out the measuring units that cancel out. Lastly, we multiply/divide the numbers involved.

$$56 \text{ cm} = 56 \text{ cm} \cdot 1 = 56 \text{ cm} \cdot \frac{1 \text{ in}}{2.54 \text{ cm}} = 56 \, \cancel{\text{cm}} \cdot \frac{1 \text{ in.}}{2.54 \, \cancel{\text{cm}}} = \frac{56 \cdot 1 \text{ in}}{2.54} = 22.047 \text{ in} \approx 22 \text{ in.}$$

    Multiply the   Replace that 1   Cancel out the   Calculate.   Round.
    quantity by 1.   with a ratio.   cm units.

Notice that we **keep the the units of measure** in the calculation! The "cm" units cancel out, and we end up with only the unit "in" (which is what we wanted: to convert the given quantity into *inches*).

Another example, of converting 8.9 quarts into liters:

$$8.9 \text{ qt} = 8.9 \text{ qt} \cdot 1 = 8.9 \text{ qt} \cdot \frac{0.946 \text{ L}}{1 \text{ qt}} = 8.9 \, \cancel{\text{qt}} \cdot \frac{0.946 \text{ L}}{1 \, \cancel{\text{qt}}} = \frac{8.9 \cdot 0.946 \text{ L}}{1} = 8.4194 \text{ L} \approx 8.4 \text{ L.}$$

    Multiply the   Replace that 1   Cancel out the   Calculate.   Round.
    quantity by 1.   with a ratio.   qt units.

2. Use the given ratios to convert the measuring units. Round your answers to one decimal digit.

a. Use $1 = \dfrac{2.54 \text{ cm}}{1 \text{ in}}$ to convert 79 inches to centimeters.

79 in =

b. Use $1 = \dfrac{1 \text{ mi}}{1.6093 \text{ km}}$ to convert 56 km to miles.

56 km =

c. Use $1 = \dfrac{1.6093 \text{ km}}{1 \text{ mi}}$ to convert 2.8 mi to kilometers.

2.8 mi =

d. Use $1 = \dfrac{0.946 \text{ L}}{1 \text{ qt}}$ to convert 4 qt to liters.

4 qt =

How do you know whether to use the ratio $\dfrac{1 \text{ in}}{2.54 \text{ cm}}$ or the ratio $\dfrac{2.54 \text{ cm}}{1 \text{ in}}$ when converting 7 inches into centimeters?

If the quantity you start with has inches, then you will need to cancel out the unit "inches" in the conversion. Therefore, choose the ratio that has inches <u>in the denominator</u>.

Here is an example of using the *wrong* ratio:

$$7 \text{ in} = 7 \text{ in} \cdot 1 = 7 \text{ in} \cdot \dfrac{1 \text{ in}}{2.54 \text{ cm}} = 7 \text{ in} \cdot \dfrac{1 \text{ in}}{2.54 \text{ cm}} = \dfrac{7 \text{ in} \cdot 1 \text{ in}}{2.54 \text{ cm}} = 2.7559 \text{ in}^2 / \text{cm}$$

| | Replace 1 with a ratio. | Nothing cancels. | Calculate. | The answer is not reasonable. Since inches are the longer units, 7 inches should convert to a bigger number of cm. The units didn't work out, either. |

Here are some conversion factors you will need in the following problems.

| 1 inch = 2.54 cm | 1 yard = 0.9144 m | 1 quart = 0.946 L | 1 lb = 0.454 kg |
| 1 foot = 0.3048 m | 1 mile = 1.6093 km | 1 ounce = 28.35 g | 1 kg = 2.2 lb |

3. Use ratios to convert the measuring units. Round your answers to one decimal digit.

a. 89 cm into inches

b. 15 kg into pounds

c. 78 miles into km

d. 89 feet into meters

e. 365 g into ounces

Chaining (optional). We can use TWO (or more) ratios in the conversion, and "chain" them together.

Example. Convert 0.9 liters into liquid ounces.

We have TWO conversion factors: 1 quart = 0.946 L and 1 quart = 32 oz. From these, we can write *four* ratios: $\frac{1 \text{ qt}}{0.946 \text{ L}}$, $\frac{0.946 \text{ L}}{1 \text{ qt}}$, $\frac{32 \text{ oz}}{1 \text{ qt}}$, and $\frac{32 \text{ oz}}{1 \text{ qt}}$, all equaling 1. We can use TWO of those four, "chaining" them together, to go from 0.9 liters to however many ounces:

$$0.9 \text{ L} = 0.9 \text{ L} \cdot \frac{1 \text{ qt}}{0.946 \text{ L}} \cdot \frac{32 \text{ oz}}{1 \text{ qt}} = 0.9 \cancel{\text{L}} \cdot \frac{1 \cancel{\text{qt}}}{0.946 \cancel{\text{L}}} \cdot \frac{32 \text{ oz}}{1 \cancel{\text{qt}}} = \frac{0.9 \cdot 32 \text{ oz}}{0.946} = \approx 30.4 \text{ oz.}$$

                      Write the two ratios   Cancel out the   Calculate.   Round.
                        that equal 1.   liters and quarts.

How do you choose which two of the possible four ratios to use? Since you start out with LITERS, you want a ratio where LITERS are in the denominator. And since you want to end up with OUNCES, you want a ratio where OUNCES are NOT in the denominator. The quarts and liters cancel out in the process, leaving the ounces.

4. Convert the measuring units as indicated.

a. Use the ratios (2.54 cm/1 in) and (12 in /1 ft) to convert 5 ft into centimeters. Round to the nearest cm.

5 ft =

b. Use the ratios (1 qt/32 oz) and (0.946 L/1 qt) to convert 24 oz into liters. Round to two decimals.

c. Convert 700 yards into meters. Round to one decimal.

d. Convert 8 kg into ounces (weight). Round to the nearest ounce.

e. Convert 371 ounces into grams. Round to the nearest 100 grams.

f. Convert 15 pints into liters. Round to two decimals.

Maps

Just like floor plans, maps also include a scale. A scale on a map may show how units on the map correspond to units in reality (for example 1 cm = 50 km). It can also be given as a ratio such as 1:120,000.

A scale of 1:120,000 means that 1 unit on the map corresponds to 120,000 units in reality. This holds true—whether you use centimeters, millimeters, or inches—because the scale 1:120,000 is a ratio without any particular unit. So 1 cm on the map corresponds to 120,000 cm in reality, and 1 inch on the map corresponds to 120,000 inches in reality.

Example 1. A map has a scale 1:150,000. How long in reality is a distance of 7.1 cm on the map?

Below you can read two solutions to this problem. Both are actually very similar!

Multiply, then change the units.

If 1 cm corresponds to 150,000 cm, then 7.1 cm corresponds to 7.1 · 150,000 cm = 1,065,000 cm.

To be useful, this figure needs to be converted into kilometers. You can do this in two steps:

1. From centimeters to meters: Since 1 m = 100 cm, we remove two zeros from 1,065,000 cm to get 10,650 meters (or you can think of it as dividing by 100).
2. From meters to kilometers: Since 1 km = 1,000 m, the 10,650 meters corresponds to 10.65 km ≈ 11 km.

Change the units, then multiply.

In this solution, we will first rewrite the scale and then use multiplication to calculate the distance in reality.

Since 1 cm corresponds to 150,000 cm, and 150,000 cm = 1,500 m = 1.5 km, we can rewrite the scale of this map as 1 cm = 1.5 km.

Then, 7.1 cm corresponds to 7.1 · 1.5 km = 10.65 km ≈ 11 km.

You can use a calculator for all the problems in this lesson.

1. A map has a scale ratio of 1:20,000. Fill in the table.

on map (cm)	in reality (cm)	in reality (m)	in reality (km)
1 cm	20,000 cm		
3 cm			
5.2 cm			
0.8 cm			
17.1 cm			

2. A map has a scale of 1:100,000.

   a. The scale says that 1 cm on the map corresponds to 100,000 cm in reality. How many kilometers is that?

   Thus, we can rewrite this scale in the format 1 cm = _____ km

   b. A ski trail measures 5.2 cm on this map. In reality, how long is the trail in kilometers?

3. A map has a scale of 1:25,000.

   a. Rewrite this scale in the format 1 cm = _____ m.

   b. Fill in the table. Give your answers to the nearest tenth of a centimeter.

on the map (cm)	in reality
	500 m
	900 m
	1.6 km
	2.5 km

4. Measure the aerial distances between the given places in centimeters and then calculate the distances in reality to the nearest kilometer. The places are marked with squares on the map. *Aerial distances* are "as the crow flies": measure them directly from point to point, not by following the roads.

   a. From Elkmont to the the Gatlinburg Welcome Center.

   b. From the Great Smoky Mountains Institute at Tremont to the Little Greenbrier School.

   c. From the Little Greenbrier School to Elkmont.

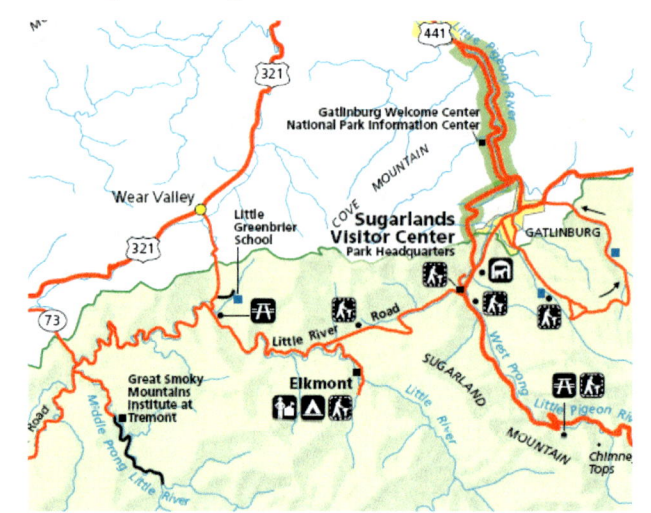

   Scale 1:180,000

5. Hannah is making a map of a farmhouse and the surrounding buildings. When she measured the distance from the farmhouse to the barn, it was 75 meters.

   a. On the map, what will the distance be from the house to the barn at a scale of 1:500?

   b. What would the distance be on a map at a scale of 1:1200?

Example 2. The distance from Jane's home to her grandma's is 220 km.
How long is the representation of that distance on a map with a scale of 1:1,500,000?

Study both solutions below, and make sure you understand them.

Divide, then convert.	Convert, then divide.
This conversion goes the other way: from reality to the map. Therefore, we need to *divide* the distance 220 km by the factor 1,500,000. We will get a very small number, and it is in kilometers just like the original distance is: $$220 \text{ km} \div 1,500,000 = 0.00014\overline{6} \text{ km}$$ However, the answer in this format is not very useful. We need to convert it into units that can be measured on a map, such as centimeters (millimeters would work, too). You could convert $0.00014\overline{6}$ km into centimeters directly, but here we will do it in two steps because that is easier for most people. (1) Converting from kilometers to meters requires multiplying our number by the unit ratio 1,000 m/km (not by 1 km/1000 m because then the units "km" wouldn't cancel): $$0.00014\overline{6} \text{ km} \cdot \frac{1,000 \text{ m}}{\text{km}} = 0.14\overline{6} \text{ m}$$ (2) Finally, we convert meters into centimeters by multiplying by 100 cm/m. We get $$0.14\overline{6} \text{ m} \cdot \frac{100 \text{ cm}}{\text{m}} = 14.\overline{6} \text{ cm.}$$ So the distance on the map is about 14.7 cm.	We will first rewrite the scale and then use division to calculate the distance in reality. According to the scale, 1 cm corresponds to 1,500,000 cm. Now, 1,500,000 cm = 15,000 m (think of dropping two zeros) and that equals 15 km (think of dropping three zeros). So we can rewrite the scale of this map as 1 cm = 15 km. Thus 220 km corresponds to 220 km/15 km = $14.\overline{6} \approx 14.7$ cm.

6. Mark is planning a route for a footrace that will be 1.5 km long. He has two city maps available. One has a scale of 1:15,000 and the other has a scale of 1:20,000. Calculate the distance of the race on each of the two maps to the nearest tenth of a centimeter.

Example 3. A map has a scale of 1:500,000. The distance from one town to another measures 5 1/4 inches on the map. How long is the distance in reality?

Again, there are two ways to solve this problem:

Multiply, then convert: Multiply the given distance 5.25 in by 500,000 and then convert the result into miles.

Convert, then multiply: Rewrite the scale into an "easier" format, then use multiplication.

Multiply, then convert.

We simply multiply 5.25 in by 500,000 to get 5.25 in · 500,000 = 2,625,000 inches. So that is the distance in reality. Next we convert this distance into miles in two steps:

(1) <u>From inches to feet</u>: Since 1 ft = 12 in, we multiply 2,625,000 inches by the ratio 1 ft/12 in (not by 12 in/1 ft because we want the inches to cancel):

$$2{,}625{,}000 \text{ in} \cdot \frac{1 \text{ ft}}{12 \text{ in}} = 218{,}750 \text{ ft}$$

You end up dividing 2,625,000 by 12, which makes sense since we are going from smaller units (inches) to bigger ones (feet), and thus we should get *fewer* units in feet.

(2) <u>From feet to miles</u>: 1 mi = 5,280 ft. Again, we multiply by the conversion ratio 1 mi/5,280 ft:

$$218{,}750 \text{ ft} \cdot \frac{1 \text{ mi}}{5{,}280 \text{ ft}} = 41.4299\overline{24} \text{ mi} \approx 41 \text{ miles}.$$

(Essentially, you divide by 5,280.)

The distance between the towns is about 41 miles.

Convert, then multiply.

The scale 1:500,000 means that 1 inch corresponds to 500,000 inches. We need to convert that to a more useful unit, such as feet or miles. Similar to above, the conversions go like this:

$$500{,}000 \text{ in} \cdot \frac{1 \text{ ft}}{12 \text{ in}} = 41{,}666.\overline{6} \text{ ft} \quad \text{and} \quad 41{,}666.\overline{6} \text{ ft} \cdot \frac{1 \text{ mi}}{5{,}280 \text{ ft}} = 7.89\overline{14} \text{ mi} \approx 7.90 \text{ miles}$$

So the scale of our map is 1 inch = 7.90 miles.

Then, the given distance 5.25 miles corresponds to 5.25 · 7.90 miles ≈ 41 miles.

Either way, the most difficult part is in converting inches into miles. Since we are dealing with customary units, there aren't any shortcuts that would allow us to convert the measuring units without a calculator, so neither solution is really easier than the other.

7. A map has a scale ratio of 1:400,000. In miles, how long is a nature hike that measures 2.5 inches on the map? Give your answer to the nearest mile.

8. Use a map you have on hand, and measure distances on it with a ruler. Then calculate the distances in reality and give them to a reasonable accuracy. If you don't have a map on hand, skip this exercise and just do the next one.

9. On this map of the USA measure the distances in inches. Then calculate the distances in reality and give them to the nearest hundred miles.

   a. From Tallahassee to Denver.

   b. From Sacramento to Austin.

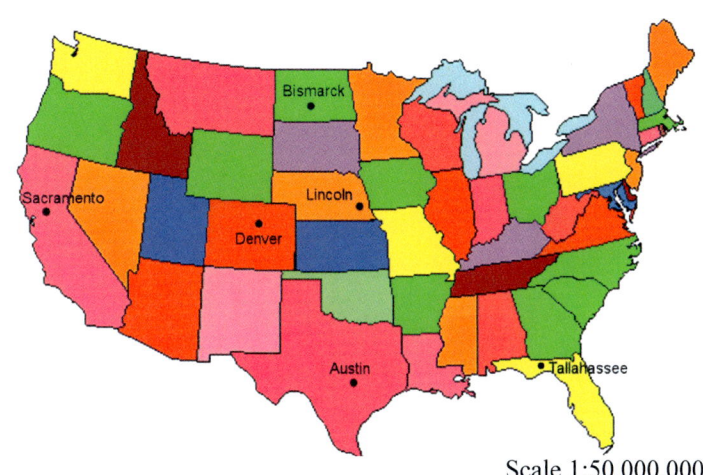

   Scale 1:50,000,000

   c. From Lincoln to Bismarck.

10. An island is 16.2 miles from the mainland. What is that distance on a map with a scale of 1:500,000? Finish Ellie's solution to this problem by filling in the words *multiply* and *divide*, the sign "·" or "÷," and numbers. Give your final answer to the tenth of a inch.

    First, I _____ the distance 16.2 miles by the factor 500,000. I will get a very small number, which will be in miles: 16.2 miles ▢ 500,000 = _____ miles.

    Next I convert this to feet, and then to inches.

    Converting miles to feet means to _____ by the ratio 5,280 ft/1 mi:

    _____ · $\dfrac{5{,}280 \text{ ft}}{1 \text{ mi}}$ = _____

    Then I convert the result from feet to inches by _____ing by the ratio 12 in/ 1 ft:

    _____ · $\dfrac{12 \text{ in.}}{1 \text{ ft}}$ = _____ in. ≈ _____ in.

11. The distance from Mark's home to the airport is 45.62 miles according to an online distance calculator.
    How long would this distance be, in inches, on a map with a scale of 1:250,000?
    How about on a map with a scale of 1:300,000?

12. The scale of a map is 1:15,000. A rectangular plot of land measures 1 3/16″ by 2 1/8″ on the map.
    a. Find the area of the land in reality in square feet. Don't round your answer, as we will use the answer in part (b).

    b. Calculate the area of the land in acres, to the nearest tenth of an acre.
    Use 1 acre = 43,560 square feet.

13. The length of a hiking path is 5.0 inches on a map with a scale of 1:200,000.
    What would the length be on a map with a scale of 1:150,000?

Puzzle Corner

A sheet of A4 paper measures 210 mm by 297 mm. You want to print a map of a plot of land with the dimensions of 1.65 km by 2.42 km onto one sheet of A4 paper. What scale should you use for your map so that it fits onto the sheet of A4 paper?

Significant Digits

(This lesson is optional.)

Example 1. In reality, how long is a distance of 5.7 cm on a map with a scale of 1:400,000?

Since 1 cm corresponds to 400,000 cm, then 5.7 cm corresponds to 5.7 · 400,000 cm = 2,280,000 cm. Converting this into kilometers we get 22.8 km.

However, since our measurement was only to the accuracy of a tenth of a centimeter, we cannot truthfully give our answer to an accuracy of 22.8 km. You see, the measurement 5.7 cm is an *approximation*. The true distance on the map could be 5.7352 cm or 5.67364 cm — we don't know since we cannot measure it that accurately.

Let's consider some other distances on the map that would be rounded to 5.7 cm, and calculate them in reality. Study the table on the right:

5.65 cm on map = 22.6 km in reality
5.688 cm on map = 22.752 km in reality
5.703 cm on map = 22.812 km in reality
5.718 cm on map = 22.872 km in reality
5.749 cm on map = 22.996 km in reality

From the table we can see that the distance in reality is anywhere from 22.6 km to about 23 km. We definitely cannot say it is exactly 22.8 km. That is why we need to round 22.8 km *to the nearest kilometer*. The distance in reality is about 23 km.

Significant digits of a number are those digits whose value contributes to the precision of the number. Significant digits help us know how to round answers when calculating *measurements*, because measurements by their nature are never totally precise.

For example, all the individual digits of 12.593 m tell us something about its precision: it is precise to the thousandth of a meter. However, in the measurement 2,000 m, we cannot be sure if the number was originally measured as 1,9283.4 m and rounded to 2,000 m or measured as 2,400 m and rounded to 2,000 m. So in 2,000 m, only the 2 is a significant digit that tells us something about its precision.

All non-zero digits are always significant. With zeros, the situation is more complex. Here are the rules:

1. All non-zero digits are significant: 38.2 has three significant digits.
2. Zeros between other significant digits are also significant: 50,039 has five significant digits.
3. Non-decimal zeros at the end of a number are not significant: 6,400 has two significant digits.
4. Decimal zeros in front of the number are not significant: 0.0038 has two significant digits.
5. Decimal zeros at the end of a number *are* significant: 0.00380 has three significant digits.

In a calculation involving multiplication and/or division, the amount of significant digits in the answer should equal the amount of significant digits in the number that is the least precise (that has the smallest amount of significant digits).

For example, 2.3 cm has two significant digits and 11.9 cm has three. When we multiply them (to get an area), we get 27.37 cm^2, but we need to take the result to only *two* significant digits (because 2.3 cm had the least amount of significant digits, which was 2), so 27.37 cm^2 gets rounded to 27 cm^2.

In this lesson you will often multiply or divide a measurement result by a conversion factor. In this situation, **keep the same number of significant digits in your converted result as what you had in your measurement.** That is because the conversion factors are more exact and have more significant digits than your measurement result, so the measurement will automatically be the number with the least amount of significant digits.

You can use a calculator for all the problems in this lesson.

1. How many significant digits do these numbers have?

a. 24.5 km	b. 20.5 km	c. 24.50 km	d. 0.5 mi
e. 15,000 ft	f. 15,001 ft	g. 0.078 km	h. 0.0780 km
i. 5,002.90 kg	j. 340 lb	k. 340.9 lb	l. 0.005 lb

2. The two sides of a rectangular play area are measured to be 24.5 m and 13.8 m.

   a. Calculate its area and give it with a reasonable amount of significant digits.

   b. Let's say the dimensions of the play area were measured more accurately to be 24.56 m and 13.89 m. Calculate the area and give the result to a reasonable accuracy.

3. Calculate the following distances in reality. Consider how many significant digits your answer should have. Note: All digits in the scale ratios are significant. For example, the scale ratio of 1:50,000 is precise to all 5 digits. It's neither 1:49,999 nor 1:50,001, but exactly 1:50,000.

   a. 6.2 cm on a map with a scale of 1:50,000

   b. 12.5 cm on a map with a scale of 1:200,000

   c. 0.8 cm on a map with a scale of 1:15,000

4. A field measures 5.0 cm by 3.5 cm on a map with a scale of 1:8,000. Calculate its area in reality.

5. The distance from Mary's home to school is 3.0 inches on a map with a scale of 1:10,000.

   a. How long is this distance in reality? Give your answer in miles to two significant digits.

   b. Give your answer in yards to two significant digits.

6. A gas station is on a rectangular plot of land that measures 45.0 m by 31.2 m. What are these dimensions on a map with a scale of 1:500?

44

Review

1. Jim cut seven 0.56-meter pieces out of a 4-meter board. How much is left?

2. Convert.

a. 0.9 m = _____ cm	**b.** 0.6 L = _____ ml	**c.** 2.2 kg = _____ g
45 cm = _____ m	5,694 ml = _____ L	390 g = _____ kg
1.5 km = _____ m	0.09 L = _____ ml	0.02 kg = _____ g

3. Convert.

a. 6 ft 11 in. = _____ in.	**b.** 2 gal = _____ C	**c.** 78 oz = ___ lb _____ oz
3 lb 11 oz = _____ oz	5 qt = _____ pt	39 in = ___ ft _____ in
3 C = _____ oz	54 oz = ___ C ___ oz	102 in = ___ ft _____ in

4. One yard is 0.9144 meters. Which is a better deal:
   40 yards of rope for $15.99
   or 100 meters of rope for $40?

5. Fill in the entries missing from this table.

Prefix	Meaning	Units - length	Units - mass	Units - volume
			centigram (cg)	
deci-				deciliter (dl)
	ten = 10		decagram (dag)	
				hectoliter (hl)

6. Change into the basic unit (meter, liter, or gram). Think of the meaning of the prefix.

   a. 34 dl   **b.** 89 cg   **c.** 16 kl

7. Convert. Use a calculator, but only in this problem!

a. 2.65 mi = _____ ft	**b.** 3,800 ft = _____ mi	**c.** 4.54 lb = _____ oz
10.9 mi = _____ yd	3,500 yd = _____ mi	10.2 ft = _____ in

8. A town map has a scale of 1:45,000.

   a. A street in this town is 850 m long. How long is that street on this map?

   b. How long in reality is a road that measures 5.4 cm on the map?

9. How many significant digits do these numbers have?

a. 270 km	b. 60.2 km	c. 20.50 km	d. 0.7 mi
e. 18,030 ft	f. 14,701 ft	g. 3.078 km	h. 0.0785 km
i. 8,042.05 kg	j. 300 lb	k. 0.99 lb	l. 0.003 lb

10. Convert the measurements into the given units.

    a. 2.7 L = _____ dl = _____ cl = _____ ml

    b. 5,600 m = _____ km = _____ dm = _____ cm

    c. 676 g = _____ dg = _____ cg = _____ mg

11. Use ratios to convert the measuring units. 1 kg = 2.2 lb, and 1 ft = 30.48 cm. Round to one decimal digit.

    a. 134 lb into kilograms

    b. 156 cm into feet

12. You have eleven empty pop bottles. Six are 350 ml, two are 2 liters, and three are 9 dl. What is the total amount of water that you can put into them?

13. Convert into the given units. Round your answers to 2 decimals if needed.

| a. 56 oz = _____ lb | c. 2.7 gal = _____ qt | e. 0.48 mi = _____ ft |
| b. 134 in = _____ ft | d. 0.391 lb = _____ oz | f. 2.45 ft = _____ ft _____ in |

Measurement Unit Conversions - Answer Key

Decimals in Measuring Units and More, p. 11

1.

a. 0.8 m = 80 cm 1.3 m = 130 cm 8.27 m = 827 cm	b. 56 cm = 0.56 m 3 cm = 0.03 m 382 cm = 3.82 m	c. 0.31 m = 31 cm 4.6 m = 460 cm 16.08 m = 1608 cm

2.

a. 0.5 km = 500 m 0.7 km = 700 m 4.5 km = 4,500 m	b. 2,400 m = 2.4 km 12,680 m = 12.68 km 540 m = 0.54 km	c. 2.001 km = 2,001 m 0.319 km = 319 m 0.04 km = 40 m

3. Jake is 1.88 m − 0.16 m = 1.72 m or 172 cm tall.

4. 4 × 550 m = 2,200 m = 2.2 km

5. (2,400 + 350) × 2 = 5,500 m or 5.5 km

6.

a. 0.7 L = 700 ml 12.6 L = 12,600 ml 0.06 L = 60 ml	b. 3,900 ml = 3.9 L 2,080 ml = 2.08 L 212 ml = 0.212 L	c. 0.009 L = 9 ml 1.35 L = 1,350 ml 1.585 L = 1,585 ml

7. The difference is 520 ml − 470 ml = 50 ml.

Teaching box:

- 0.1 kg is 100 g
- 0.01 kg is 10 g
- 0.001 kg is 1 g

8.

a. 0.3 kg = 300 g 2.6 kg = 2,600 g 0.05 kg = 50 g	b. 20 g = 0.02 kg 800 g = 0.8 kg 6,030 g = 6.03 kg	c. 1.1 kg = 1,100 g 0.152 kg = 152 g 2.093 kg = 2,093 g

9. The apple weighed 180 g. 3/4 of 180 g is 180 g ÷ 4 × 3 = 135 g.

10. 30 × 450 g = 13,500 g = 13.5 kg

11. a. 0.2 mi = 1,056 ft   b. 1.35 mi = 7,128 ft
    c. 2.7 mi = 4,752 yd   d. 0.72 mi = 1,267 yd

12. 0.7 mi + 0.65 mi + 0.5 mi × 3 = 2.85 mi = 15,048 ft

13.

a. 0.65 mi = 3,432 ft 1.34 mi = 7,075 ft	b. 0.9 mi = 1,584 yd 5.413 mi = 9,527 yd	c. 5.428 mi = 28,660 ft 2.75 mi = 4,840 yd

14. You need 55 McKinley mountains stacked on top of each other. Convert either Mt. McKinley's height into miles or the distance 211.3 miles into feet before calculating. For example, Mt. McKinley is 20,320 ÷ 5,280 = 3.8484 miles tall. Then, divide to find how many of them are needed to reach the height of the International Space Station: 211.3 mi ÷ 3.8484 = 54.905 ≈ 55.

Decimals in Measuring Units and More, cont.

15.

| a. $0.7 M = $700,000
  $2.5 M = $2,500,000 | b. $0.04 M = $40,000
  $0.39 M = $390,000 | c. $10.9 M = $10,900,000
  $2.78 M = $2,780,000 |

16. They spent $2.85 M − $0.35 M = $2.5 M = $2,500,000.

17. The men's record advanced 8.95 m − 8.21 m = 0.74 m = 74 cm. The women's record advanced 7.52 m − 6.40 m = 1.12 m = 112 cm. The women's record has advanced 38 cm more than the men's.

18. (900 cm − 2 × 90 cm) ÷ 55 cm = 13.0909. You can place 13 chairs in the row.

Rounding and Estimating, p. 15

1.

| a. 8.19 m ≈ 8 m;
  rounding error = 0.19 m | b. 362 cm ≈ 4 m;
  rounding error = 0.38 m | c. 417 cm ≈ 4 m;
  rounding error = 0.17 m |
| d. 1 m 54 cm ≈ 2 m;
  rounding error = 0.46 m | e. 14.208 m ≈ 14 m;
  rounding error = 0.208 m | f. 8 m 9 cm ≈ 8 m;
  rounding error = 0.09 m |

2.

| a. 602 m ≈ 1 km;
  rounding error = 0.398 km | b. 10.189 km ≈ 10 km;
  rounding error = 0.189 km | c. 8.057 km ≈ 8 km;
  rounding error = 0.057 km |
| d. 2,643 m ≈ 3 km;
  rounding error = 0.357 km | e. 6 km 55 m ≈ 6 km;
  rounding error = 0.055 km | f. 3,288 m ≈ 3 km;
  rounding error = 0.288 km |

3. a. 1.5 m − 0.67 m = 0.83 m
   ≈ 0.8 m
   b. 6.08 m + 0.45 m + 1.2 m = 7.73 m
   ≈ 7.7 m
   c. 1.08 m + 2.55 m = 3.63 m
   ≈ 3.6 m

4. a. 2,100 m − 293 m = 1,807 m
   ≈ 1,810 m
   b. 6,070 m + 452 m = 6,522 m
   ≈ 6,520 m
   c. 2,075 m + 3,800 m = 5,875 m
   ≈ 5,880 m

5.

| a. Estimation: 4 ÷ 0.80 = 5 books
  Exact answer: 4 ÷ 0.78 = 5.13,
  so 5 books will fit. | b. Change to grams.
  Estimation: 400 g ÷ 25 g = 16 papers
  Exact answer: 400 g ÷ 24 g = 16.6, so you
  can mail 16 papers. |

6. Answers will vary. For example: 18 × 2.2 cm ≈ 20 × 2 cm = 40 cm. It is better to round both factors, one up to 20, and the other down to 2 cm, than to only round 2.2 cm to 2 cm and leave the 18 without rounding it. You can see that by checking the exact answer: it is 39.6 cm, which is really close to our estimate of 40 cm.

7. Answers will vary. Notice we need to change 146 cm into meters. For example: 2 ½ × 1.46 m ≈ 3 × 1.5 m = 4.5 cm. Or, 2 ½ × 1.46 m ≈ 3 × 1.4 m = 4.2 m. Or, 2 ½ × 1.46 m ≈ 2 ½ × 1.5 m = 3 m + 0.75 m = 3.75 m. The last estimation is the most accurate.

8. He jogs 3 × 1,425 m = 4,275 meters, which is about 4,300 meters.

9. a. $50 − (10 × $2.46)

   b. |⎯⎯⎯⎯⎯⎯⎯⎯⎯|
      ⎯⎯⎯ $ 50 ⎯⎯⎯

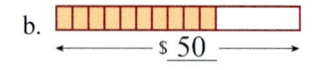

   c. He has about $50 − (10 × $2.50) = $25 left.
   d. He has $50 − (10 × $2.46) = $25.40 left.

48

The Metric System, p. 17

1.

a. 2 cm = 2/100 m = 0.02 m 6 dm = 6/10 m = 0.6 m 8 mm = 8/1000 m = 0.008 m	b. 3 dam = 30 m 9 km = 9,000 m 2 hm = 200 m	c. 6 mm = 0.006 m 20 cm = 0.20 m 8 dm = 0.8 m

2.

a. 2 ml = 2/1000 L = 0.002 L 6 cl = 6/100 L = 0.06 L 8 dg = 8/10 g = 0.8 g	b. 7 dl = 0.7 L 6 mg = 0.006 g 8 dl = 0.8 L	c. 3 dag = 30 g 8 kg = 8,000 g 2 hl = 200 L

3.

a. 75.4 m

km	hm	dam	m	dm	cm	mm
		7	5.	4		

c. 4.6 km

km	hm	dam	m	dm	cm	mm
4.	6					

b. 843 mm

km	hm	dam	m	dm	cm	mm
			8	4	3	

d. 35.49 dam

km	hm	dam	m	dm	cm	mm
	3	5.	4	9		

4.

	to m	to dm	to cm	to mm
a. 75.4 m	75.4	754	7,540	75,400
b. 843 mm	0.843	8.43	84.3	843

5.

	to hm	to dam	to m	to dm
a. 4.6 km	46	460	4,600	46,000
b. 35.49 dam	3.549	35.49	354.9	3,549

6. a. hectograms   b. centigrams   c. decigrams

7. a. 4,500 dl   b. 450.0 L (or 450 L)   c. 45.00 dal (or 45 dal)   d. 4.500 hl (or 4.5 hl)

8.

	a. 5,000 mm	b. 380 cm	c. 6.5 dm
meters	5 m	3.8 m	0.65 m
decimeters	50 dm	38 dm	6.5 dm
centimeters	500 cm	380 cm	65 cm
millimeters	5,000 mm	3,800 mm	650 mm

9. It will last 10 days. 200 ml is equal to 20 cl.

10. a. Hannah 151 cm; Erica 136 cm   b. Hannah 175 cm; Erica 160 cm   c. Hannah 165 cm; Erica 150 cm

11. a. 200 × 14 dg = 2,800 dg = 280 g.   b. four boxes

Convert Metric Measuring Units, p. 20

1.

a. 3 cm = 3/100 m = 0.03 m	b. 2 cg = 2/100 g = 0.02 g
5 mm = 5/1000 m = 0.005 m	6 ml = 6/1000 L = 0.006 L
7 dl = 7/10 L = 0.7 L	1 dg = 1/10 g = 0.1 g

2.

a. 3 kl = 3,000 L	b. 2 dam = 20 m	c. 70 km = 70,000 m
8 dag = 80 g	9 hl = 900 L	5 hg = 500 g
6 hm = 600 m	7 kg = 7,000 g	8 dal = 80 L

3.

a. 3,000 g = 3 kg	b. 0.01 m = 1 cm	c. 0.04 L = 4 cl
800 L = 8 hl	0.2 L = 2 dl	0.8 m = 8 dm
60 m = 6 dam	0.005 g = 5 mg	0.007 L = 7 ml

4. a. 0.04 meters = 4 cm   b. 0.005 grams = 5 mg   c. 0.037 meters = 37 mm
   d. 400 liters = 4 hl   e. 0.6 meters = 6 dm   f. 2,000 meters = 2 km
   g. 0.206 liters = 206 ml   h. 20 meters = 2 dam   i. 0.9 grams = 9 dg

5. a. 45 cm = 0.45 m   b. 65 mg = 0.065 g   c. 2 dm = 0.2 m
   d. 81 km = 81,000 m   e. 6 ml = 0.006 L   f. 758 mg = 0.758 g
   g. 2 kl = 2,000 L   h. 8 dl = 0.8 L   i. 9 dag = 90 g

6. a. 12.3 m

km	hm	dam	m	dm	cm	mm
		1	2.	3		

c. 56 cl

kl	hl	dal	l	dl	cl	ml
			0.	5	6	

b. 78 mm

km	hm	dam	m	dm	cm	mm
			0.	0	7	8

d. 9.83 hg

kg	hg	dag	g	dg	cg	mg
	9	8	3.			

7.

	m	dm	cm	mm
a. 12.3 m	12.3	123	1230	12300
b. 78 mm	0.078	0.78	7.8	78
	L	dl	cl	ml
c. 56 cl	0.56	5.6	56	560
	g	dg	cg	mg
d. 9.83 hg	983	9830	98300	983000

Convert Metric Measuring Units, cont.

8. a. 560 cl = 5.6 L
   c. 24.5 hm = 245,000 cm
   e. 35,200 mg = 35.2 g
   g. 0.483 km = 4830 dm
   i. 1.98 hl = 1980 dl

   b. 0.493 kg = 49.3 dag
   d. 491 cm = 4.91 m
   f. 32 dal = 32,000 cl
   h. 0.0056 km = 560 cm
   j. 9.5 dl = 0.95 L

9. a. 13 cm   b. 45 mm
   c. 0.92 m   d. 2.4 m
   e. 1.70 m   f. 1.34 m

10. a. The books weigh 1200 g + 1040 g + 520 g + 128 g = 2,888 g, or 2.888 kg.
    b. The total volume of the containers is 1.4 L + 2.25 L + 0.55 L + 0.24 L + 0.4 L = 4.84 L.

11. You can fill the 4-ml dropper 0.200 ÷ 0.004 = 50 times from the 2-dl bottle.

12. The patient will have received 2 grams of medicine <u>in ten days</u>. The 70-kg patient receives 3 mg/kg × 70 kg = 210 mg per day. Two grams is 2,000 mg. Nine days is not quite enough time, since 9 × 210 mg = 1,890 mg. So it is the tenth day that the patient finishes receiving the 2 g of medicine.

Converting Between Customary Units of Measurement, p. 23

1.

a. 6 ft = 72 in 7 ft 5 in. = 89 in	b. 25 in = 2 ft 1 in 45 in = 3 ft 9 in	c. 13 ft 7 in = 163 in 71 in. = 5 ft 11 in

2.

a. 2 lb 8 oz = 40 oz 45 oz = 2 lb 13 oz	b. 8 lb = 128 oz 56 oz = 3 lb 8 oz	c. 43 oz = 2 lb 11 oz 90 oz = 5 lb 10 oz

3.

a. 3 C = 24 oz 55 oz = 6 C 7 oz	b. 4 C = 2 pt 3 pt = 6 C	c. 7 gal = 28 qt 45 qt = 11 gal 1 qt

4. a. Multiply: 11 yd × 3 × 12 = 396 in.
   b. Divide: 711 ÷ 12 = 59 R3, so 711 in. = 59 ft 3 in.
   c. Divide: 982 ÷ 12 = 81 R10, so 982 in. = 81 ft 10 in. Then divide 81 ft by 3 to get 27 yd. So, 982 in. = 27 yd 0 ft 10 in.
   d. First, divide by eight: 254 ÷ 8 = 31 R6, so 254 oz = 31 C 6 oz.
      Next, convert the 31 cups into quarts and cups: 31 C ÷ 4 = 7 qt 3 C. We get 254 oz = 7 qt 3 C 6 oz.
      Lastly, write 7 qt as 1 gal 3 qt., and get 254 oz = 1 gal 3 qt 3 C 6 oz.

5.

a. 7.4 mi = 39,072 ft 16,000 ft = 3.03 mi	b. 1,500 ft = 500 yd 7,500 yd = 4.26 mi	1,760 mile mi 3 yard yd 12 foot ft inch in
c. 900 ft = 0.17 mi 2.56 mi = 4505.6 yd	d. 12.54 mi = 66211.2 ft 82,000 ft = 15.53 mi	1 mile = 5,280 feet

6.

a. 15.2 lb = 243.2 oz 655 oz = 40.94 lb	b. 4.78 T = 9,560 lb 7,550 lb = 3.78 T	c. 78 oz = 4.88 lb 0.702 T = 1,404 lb

Converting Between Customary Units of Measurement, cont.

7.
F 0.6 mi = 3,168 ft	**G** 7 C = 56 oz	**I** 14,256 ft = 2.7 mi
A 5,632 yd = 3.2 mi	**R** 6,200 lb = 3.1 T	**W** 6 ft 7 in = 79 in
O 10 qt = 40 C	**S** 3 lb 5 oz = 53 oz	**L** 732 in = 61 ft
H 2 lb 11 oz = 43 oz	**E** 5 ft 2 in = 62 in	**D** 42 in = 3.5 ft
L 1.3 mi = 2,288 yd	**O** 40 oz = 2.5 lb	**P** 3 gal = 24 pt
		A 0.75 mi = 3,960 ft

What did one potato chip say to the other?

53	43	3960	61	2288		79	62		56	40
S	H	A	L	L		W	E		G	O

3168	2.5	3.1		3.2		3.5	2.7	24	
F	O	R		A		D	I	P	?

8. a. Two gallons. There are 16 cups in 1 gallon. 2 × 16 C = 32 C; you will need two *whole* gallons to have 30 cups.
   b. She made 72 jars of applesauce. One gallon is 4 quarts, so in quarts, Mom got 2 × 9 × 4 qt = 72 qt.
   c. You can cut out fourteen 8-inch pieces. 9 3/4 ft in inches is 9.75 × 12 in. = 117 in. Then, divide that by 8 in:
      117 in. ÷ 8 in = 14 R5, so you can cut fourteen 8-inch pieces and you'll have 5 inches left over.
   d. One ounce costs $1.20 ÷ 4 = $0.30, so 5 ounces cost 5 × $0.30 = 1.50.
   e. It weighs 16 lb 12 oz. In ounces, the box weighs 20 × 13 oz + 8 oz = 268 oz. To convert that to pounds, divide by 16:
      268 oz ÷ 16 oz = 16 R12. The box of shampoo weighs 16 lb 12 oz.
   f. Either 56 C 2 oz, or 14 qt 2 oz, or 3 gal 2 qt 2 oz. In a month, Mark drinks 3 × 5 oz × 30 = 450 oz. To convert that to
      cups and ounces, divide by eight: 450 ÷ 8 = 56 R2, so 450 oz = 56 C 2 oz. To further convert the 56 cups into quarts,
      divide by four: 56 ÷ 4 = 14, so 56 C = 14 qt. And, 14 qt can be written as 3 gal 2 qt.
   g. She lost 5 × 16 = 80 oz in 28 days. On average, she lost 80 ÷ 28 ≈ 2.9 ounces per day.
      Or, in pounds, she lost on average 5 lb ÷ 28 ≈ 0.18 pounds per day.

Convert Customary Measuring Units, p. 27

1. a. the upper   b. the lower   c. the lower   d. the upper

2.
a. 564 ft = 0.11 mi	c. 3,400 yd = 1.93 mi	e. 0.28 mi = 1478.4 ft
b. 45,000 ft = 8.52 mi	d. 7.8 mi = 41,184 ft	f. 10.17 mi = 17899.2 yd

3.
a. 3 in = 0.25 ft	c. 14.7 ft = 176.4 in	e. 281 in = 23.42 ft
b. 21 in = 1.75 ft	d. 0.8 ft = 9.6 in	f. 7 1/3 ft = 88 in

4.
a. 5 oz = 0.31 lb	c. 3.6 lb = 57.6 oz	e. 127 oz = 7.94 lb
b. 35 oz = 2.19 lb	d. 0.391 lb = 6.26 oz	f. 6 3/4 lb = 108 oz

5.
a. 6.4 gal = 25.6 qt	d. 0.56 qt = 17.92 fl. oz.	g. 0.054 T = 108 lb
b. 78 fl. oz. = 2.44 qt	e. 560 qt = 140 gal	h. 1,200 lb = 0.6 T
c. 2.3 qt = 73.6 fl. oz.	f. 3.2 T = 6,400 lb	i. 6,750 lb = 3.38 T

6.
a. 2 ft 6 in = 30 in	c. 162 in = 13 ft 6 in	e. 254 in = 21 ft 2 in
b. 7 ft 11 in = 95 in	d. 79 in = 6 ft 7 in	f. 1,028 in = 85 ft 8 in

7.
a. 6 lb 9 oz = 105 oz	c. 86 oz = 5 lb 6 oz	e. 483 oz = 30 lb 3 oz
b. 11 lb 12 oz = 188 oz	d. 145 oz = 9 lb 1 oz	f. 591 oz = 36 lb 15 oz

Convert Customary Measuring Units, cont.

8.

a. 2.7 ft = 2 ft 8 in	c. 3.15 ft = 3 ft 2 in	e. 55.46 lb = 55 lb 7 oz
b. 10.2 ft = 10 ft 2 in	d. 7.8 lb = 7 lb 13 oz	f. 8.204 lb = 8 lb 3 oz

9. a. 13 lb 11 oz   b. 109 ft 2 in   c. 14 qt 30 oz   d. 45 ft 6 in   e. 12 lb 3 oz   f. 2 h 49 min

10. 4 ft 3 in − (3 × 2 $\frac{3}{8}$) in = 4 ft 3 in − 7 1/8 in = 3 ft 7 $\frac{7}{8}$ in or about 3 ft 8 in.
    Jack was 3 feet 8 inches tall three years ago.

11. The room is 40 ft × 12 in/ft = 480 inches wide. Since each chair is 21 inches, and
    480 ÷ 21 ≈ 22.857, you can place 22 chairs in one row.

    The two 3-ft aisles take up 6 ft × 12 in/ft = 72 inches, so there are 480 in − 72 in = 408 inches
    left for the chairs. This time, since 408 ÷ 21 ≈ 19.429, you can fit 19 chairs in one row.

12. 4 × 32 ÷ 6 = 21.33. Jack would need 21 glasses for the tea.

13. (60 × 16) ÷ (2 × 16 + 3) = 27.43. You can pack 27 math books in the box.

14. a. $5.69 ÷ 13 = $0.437692307... per ounce $13.99 ÷ 32 = $0.4371875 per ounce
        The 1-quart bottle is the better deal.
    b. $0.35 ÷ 12 = $0.0291666... per ounce. $4.10 ÷ (8 × 16) = $0.03203125 per ounce.
        Twelve ounces for $0.35 is the better deal.

15. ($4.84 ÷ 25.5) × 32 = $6.0737. A quart of olive oil would cost $6.07.

16. a. 128 oz ÷ 17 = 7.53 oz Each person would get about 7 1/2 ounces of ice cream.
    b. Each person would get 5 scoops of ice cream.

17. (10 × 16 oz) ÷ 7 = 22.85714 oz. = 1 lb 6.85714 oz. Each person would get 1 lb 7 oz.

Convert Between Customary and Metric, p. 31

1. a. 2.5 cm   b. 30 cm   c. 1.5 km   d. 1 L
   e. 4 lb   f. 3 ft   g. 240 ml   h. 4 L

2.

a. 1 cm < 1 in	b. 1 L > 1 qt	c. 1 kg > 1 lb	d. 1 g < 1 oz
e. 4 in < 20 cm	f. 5 kg < 20 lb	g. 3 gal > 2 L	h. 7 m > 4 ft

3.

a.	b.	c.	d.
1 cm = 0.39 in 25 cm = 9.84 in	1 m = 1.09 yd 5.4 m = 17.72 ft	2 L = 2.11 qt 4.6 L = 4.86 qt	5 kg = 11.00 lb 0.568 kg = 1.25 lb
e.	f.	g.	h.
5 in = 12.70 cm 10 in = 25.40 cm	30 ft = 9.14 m 22 ft = 6.71 m	1 gal = 3.78 L 3 1/2 qt = 3.31 L	75 lb = 34.09 kg 8.5 lb = 3.86 kg

4. (c) 90 km/h. (Miles are *bigger* than kilometers, so there will be *more* kilometers. So 55 mi/hr × 1.6 km/mi = 88 km/hr.)

5. In ounces:

    The 24-oz bottle costs $6.75 ÷ 24 = $0.28125 per ounce.
    1 qt is 0.946 liters, so 1 liter = 1/0.946 = 1.05708 quarts. In ounces, this is 32 × 1.05708 = 33.826638 ounces.

    The 1-liter bottle costs $9.25 ÷ 33.826638 = $0.27345 per ounce.
    The 1-liter container is the better buy.

    In liters:

    24 ounces is 3/4 of a quart. 1 qt is 0.946 liters, so 3/4 qt = (3/4) × 0.946 L = 0.7095 liters.
    The 24-oz bottle costs $6.75 ÷ 0.7095 = $9.5137 per liter, and the 1-liter bottle costs $9.25 per liter.
    The latter is the better deal.

6. a. <u>No</u>, because 64 oz is equivalent to 2 qt, and a quart is smaller than a liter.
       64 oz = 8 C = 4 pt = 2 qt = 2 × 0.946 L = 1.892 L.

Convert Between Customary and Metric, cont.

6. b. The 64-oz container is 2.2 L − 1.892 L = <u>0.308 L less</u> than the 2.2-liter container.

7. From the lightest to the heaviest: Angela (56 kg ≈ 123 lb), Theresa (128 lb), Elizabeth (60 kg = 132 lb), and Judy (137 lb).

8. If a marathon is 26.21875 miles (*i.e.*, 26 miles and 385 yards), then a half marathon is
26.21875 mi ÷ 2 × 1.6093 km/mi = <u>21.097 km</u>, or 21 km and 97 m.

Using Ratios to Convert Measuring Units, p. 33

1.

1 ft = 0.3048 m	1 ounce = 28.35 g	1 mi = 1,760 yd	1 m = 1.0936 yd
↓	↓	↓	↓
$\dfrac{1 \text{ ft}}{0.3048 \text{ m}} = 1$	$\dfrac{1 \text{ oz}}{28.35 \text{ g}} = 1$	$\dfrac{1 \text{ mi}}{1{,}760 \text{ yd}} = 1$	$\dfrac{1 \text{ m}}{1.0936 \text{ yd}} = 1$

2.

a. 79 in = 79 in · 1 = 79 in · $\dfrac{2.54 \text{ cm}}{1 \text{ in}}$ = 79 · 2.54 cm = 200.66 cm ≈ 200.7 cm	
b. 56 km = 56 km · 1 = 56 km · $\dfrac{1 \text{ mi}}{1.6093 \text{ km}}$ = $\dfrac{56 \text{ mi}}{1.6093}$ = 34.7977 mi ≈ 34.8 mi	
c. 2.8 mi = 2.8 mi · 1 = 2.8 mi · $\dfrac{1.6093 \text{ km}}{1 \text{ mi}}$ = 2.8 · 1.6093 km = 4.50604 km ≈ 4.5 km	
d. 4 qt = 4 qt · 1 = 4 qt · $\dfrac{0.946 \text{ L}}{1 \text{ qt}}$ = 4 · 0.946 L = 3.784 L ≈ 3.8 L	

3.

a. 89 cm = 89 cm · 1 = 89 cm · $\dfrac{1 \text{ in}}{2.54 \text{ cm}}$ = $\dfrac{89 \text{ in}}{2.54}$ = 35.0394 in ≈ 35.0 in
b. 15 kg = 15 kg · 1 = 15 kg · $\dfrac{2.2 \text{ lb}}{1 \text{ kg}}$ = 15 · 2.2 lb = 33 lb
c. 78 mi = 78 mi · 1 = 78 mi · $\dfrac{1.6093 \text{ km}}{1 \text{ mi}}$ = $\dfrac{78 \cdot 1.6093 \text{ km}}{1}$ = 125.529 km ≈ 125.5 km
d. 89 ft = 89 ft · 1 = 89 ft · $\dfrac{0.3048 \text{ m}}{1 \text{ ft}}$ = 89 · 0.3048 m = 27.1272 m ≈ 27.1 m
e. 365 g = 365 g · 1 = 365 g · $\dfrac{1 \text{ oz}}{28.35 \text{ g}}$ = $\dfrac{365 \text{ oz}}{28.35}$ = 12.8748 oz ≈ 12.9 oz

4.

a. 5 ft = 5 ft · $\dfrac{2.54 \text{ cm}}{1 \text{ in}}$ · $\dfrac{12 \text{ in}}{1 \text{ ft}}$ = 5 · 12 · 2.54 cm = ≈ 152 cm
b. 24 oz = 24 oz · $\dfrac{1 \text{ qt}}{32 \text{ oz}}$ · $\dfrac{0.946 \text{ L}}{1 \text{ qt}}$ = $\dfrac{24 \cdot 0.946 \text{ L}}{32}$ = 0.7095 ≈ 0.71 L
c. 700 yd. = 700 yd · $\dfrac{0.9144 \text{ m}}{1 \text{ yd}}$ = (700 yd · 0.9144 m/1yd) = 700 · 0.9144 m = 640.08 m ≈ 640.1 m
d. 8 kg = 8 kg · $\dfrac{2.2 \text{ lb}}{1 \text{ kg}}$ · $\dfrac{16 \text{ oz}}{1 \text{ lb}}$ = 8 · 2.2 · 16 oz = 281.6 ≈ 282 oz
e. 371 oz = 371 oz · $\dfrac{28.35 \text{ g}}{1 \text{ oz}}$ = 371 · 28.35 g = 10517.85 ≈ 10,500 g
f. 15 pt = 15 pt · $\dfrac{0.946 \text{ L}}{1 \text{ qt}}$ · $\dfrac{1 \text{ qt}}{2 \text{ pt}}$ = $\dfrac{15 \cdot 0.946 \text{ L}}{2}$ = 7.095 ≈ 7.10 L

Maps, p. 37

1.

on map (cm)	in reality (cm)	in reality (m)	in reality (km)
1 cm	20,000 cm	200 m	0.2 km
3 cm	60,000 cm	600 m	0.6 km
5.2 cm	104,000 cm	1,040 m	1.04 km
0.8 cm	16,000 cm	160 m	0.16 km
17.1 cm	342,000 cm	3,420 m	3.42 km

2. a. 1 km; the scale is 1 cm = 1 km
   b. The actual length of the ski trail is 5.2 kilometers.

3. a. 1 cm is in reality 25,000 cm = 250 m = 0.25 km.
      So the scale becomes 1 cm = 250 m.
   b. See the table on the right.

on the map (cm)	in reality
2.0 cm	500 m
3.6 cm	900 m
6.4 cm	1.6 km
10.0 cm	2.5 km

4. Check the student's answers. The size of the map will vary according to the printer settings when it was printed. If the page was printed at "100% of normal size" (and not "scale to fit"), the answers should match the ones given below.

   a. The distance on the map is 3.5 cm. In reality, it is 3.5 · 180,000 cm = 630,000 cm = 6,300 m = 6.3 km.
   b. The distance on the map is 2.2 cm. In reality, it is 2.2 · 180,000 cm = 396,000 cm = 3,960 m ≈ 4.0 km.
   c. The distance on the map is 1.9 cm. In reality, it is 1.9 · 180,000 cm = 342,000 cm = 3,420 m ≈ 3.4 km.

5. a. At a scale of 1:500, 75 m is 7500 cm ÷ 500 = 15 cm on the map.
   b. At a scale of 1:1200, 75 m is 7500 cm ÷ 1200 = 6¼ cm on the map.

6. The distance 1.5 kilometers is 1,500 meters and 150,000 centimeters. Once we have converted the distance to centimeters, we can simply divide. For the map with a scale of 1:15,000, we calculate 150,000 cm ÷ 15,000 = 10.0 cm. For the map with a scale of 1:20,000, we calculate 150,000 cm ÷ 20,000 = 7.5 cm.

7. The nature hike is approximately 16 miles long.
   In reality, the hike is 2.5 in · 400,000 = 1,000,000 in = 1,000,000 in · 1 ft/(12 in) · 1 mi/(5,280 ft) = 15.7$\overline{828}$ mi.

8. Please check the student's work.

9. Check the student's answers. If the lesson was printed, the size of the map will depend on how the printer scaled the printing. If it printed the page at 100% (and not "scale to fit"), then the answers should match the answers below. If it printed it at a different size, then the given scale on the map (1:50,000,000) is not correct. Even so, you can still check the student's answers, but the answers will neither match the ones below nor the distances in reality.

   a. The distance on the map from Tallahassee to Denver is 1 11/16 in. So in reality, it is 1 11/16 in · 50,000,000
      = 27/16 · 50,000,000 in = 84,375,000 in = 84,375,000 in · (1 ft)/(12 in) · (1 mi)/(5,280 ft) ≈ 1331.68 mi
      ≈ 1,300 miles.

   b. The distance on the map from Sacramento to Austin is 1 7/8 in. In reality, it is 1 7/8 in · 50,000,000
      = 15/8 · 50,000,000 in = 93,750,000 in = 93,750,000 in · (1 ft)/(12 in) · (1 mi)/(5,280 ft) ≈ 1,479.64 mi
      ≈ 1,500 miles.

   c. The distance on the map from Lincoln to Bismarck is 9/16 in. So in reality, it is 9/16 in · 50,000,000
      = 9/16 · 50,000,000 in = 28,125,000 in = 28,125,000 in · (1 ft)/(12 in) · (1 mi)/(5,280 ft) ≈ 400 miles.

Maps, cont.

10. > First, I <u>divide</u> the distance 16.2 miles by the factor 500,000. I will get a very small number, which will be in miles: 16.2 miles ÷ 500,000 = <u>0.0000324</u> miles.
   > Next I convert this to feet, and then to inches.
   > Converting miles to feet means to <u>multiply</u> by the ratio 5,280 ft/1 mi:
   >
   > $\underline{0.0000324} \cdot \dfrac{5{,}280 \text{ ft}}{1 \text{ mi}} = \underline{0.171072}$
   >
   > Then I convert the result from feet to inches by <u>multiply</u>ing by the ratio 12 in/1 ft:
   >
   > $\underline{0.171072} \cdot \dfrac{12 \text{ in.}}{1 \text{ ft}} = \underline{2.052864}$ in ≈ <u>2.1</u> in.

11. Since we will want to work on the map in inches, let's convert 45.62 miles to inches:
    First 45.62 mi · 5280 ft/mi = 240873.6 ft, and 240873.6 ft · 12 in/ft = 2,890,483.2 in.
    On a map with a scale of 1:250,000, 45.62 miles is 2,890,483.2 in ÷ 250,000 = 11.5619328 in ≈ 11 $^9/_{16}$ in.
    On a map with a scale of 1:300,000, 45.62 miles is 2,890,483.2 in ÷ 300,000 = 9.634944 in ≈ 9 $^5/_8$ in.

    (To convert decimal inches to a fraction:
    (1) Subtract the integer part: 11.5619328 in − 11 in = 0.5619328 in; 9.634944 in − 9 in = 0.634944 in.
    (2) Multiply by 16 (Fractional inches are measured in fourths, eighths, sixteenths, *etc*. It is unlikely that you will need to or even be able to measure much more accurately than a sixteenth of an inch):
        0.5619328 in · 16 = 8.9909248; 0.634944 in · 16 = 10.159104.
    (3) Round to the nearest sixteenth: 8.9909248 ≈ 9; 10.159104 ≈ 10.
    (4) If the number of sixteenths comes out even, then convert to eighths, fourths,
        or a half as is appropriate: $^{10}/_{16} = {}^5/_8$.)

12. In decimals 1 3/16″ by 2 1/8″ is 1.1875 in by 2.125 in. In reality the dimensions are
    1.1875 in · 15,000 = 17812.5 in by 2.125 in · 15,000 = 31,875 in. Converting to feet gives
    17812.5 in · 1 ft/12 in = 1484.375 ft by 31,875 in · 1 ft/12 in = 2656.25 ft.

    a. The (unrounded) area in square feet is 1,484.375 ft · 2,656.25 ft = 3,942,871.09375 ft².

    b. The exact answer in (a) expressed in tenths of an acre would be:
       3,942,871.09375 ft² · (1 acre / 43,560 ft²) ≈ 90.5 acres.

13. We can simply multiply the distance 5.0 inches by the ratio of the two maps' scales —
    150,000/200,000 = 3/4 or 200,000/150,000 = 4/3. Since 1:150,000 is closer to reality
    than 1:200,000 is, we expect the length of the hiking trail to be longer on the map at 1:150,000.
    So we multiply the 5.0 inches by 4/3 to get 5.0 inches · 4/3 = 6 2/3 in ~ 6.7 inches. (The original
    length was given in tenths of an inch (5.0 in), so the answer should be to that same accuracy.)

    Another way to solve this is to first find the length of the trail in reality, and then find the length
    of the trail on the map at 1:150,000. In reality, the trail is 5 in · 200,000 = 1,000,000 in
    = 15.782828283 mi. On the other map, this distance will be 15.782828283 mi ÷ 150,000
    = 0.000105219 mi = 6.666666667 in.

Puzzle Corner:
Let's calculate both dimensions, then compare them. The plot of land measures 1.65 km by 2.42 km, which is 1,650 m by 2,420 m, which is 1,650,000 mm by 2,420,000 mm. Using the short sides gives a scale factor of 210 mm : 1,650,000 mm ≈ 1 : 7857. Using the long sides gives a scale factor of 297 mm : 2,420,000 mm ≈ 1 : 8148. To make the map fit onto the paper, we have to choose the smaller scale, which is the bigger number, so the answer is 1 : 8148. (In reality, we would probably want to leave a margin unprinted around the edge of the paper, especially if we were printing from an electronic printer that cannot print all the way to the edge, so we would be more likely to round the answer to 1:8500 or even 1:9000 or 1:10,000.)

Significant Digits, p. 43

1. a. 3 digits b. 3 digits c. 4 digits d. 1 digit e. 2 digits f. 5 digits
   g. 2 digits h. 3 digits i. 6 digits j. 2 digits k. 4 digits l. 1 digit

2. a. Here we need to give the answer to 3 significant digits since both dimensions are given to 3 significant digits:
   24.5 m · 13.8 m = 338.1 m² ≈ 338 m²

   b. Now both dimensions are given to 4 significant digits, so the answer is given to 4 significant digits also:
   24.56 m · 13.89 m = 341.1384 m² ≈ 341.1 m²

3. a. Since 6.2 cm has two significant digits and the scale ratio has five, we give the final answer to two significant digits:
   6.2 cm · 50,000 = 310,000 cm = 3.1 km

   b. Since 12.5 cm has three significant digits and the scale ratio has six, we give the final answer to three significant digits:
   12.5 cm · 200,000 = 2,500,000 cm = 25.0 km

   c. Since 0.8 cm has one significant digit and the scale ratio has five, we give the final answer to one significant digit:
   0.8 cm · 15,000 = 12,000 cm = 0.12 km ≈ 0.1 km

4. The dimensions 5.0 cm and 3.5 cm are given to two significant digits. Let's calculate those dimensions in reality, and give them to two significant digits:

   5.0 cm · 8,000 = 40,000 cm = 400 m = 0.40 km

   3.5 cm · 8,000 = 28,000 cm = 280 m = 0.28 km

   The area needs also be given to two significant digits, since both numbers we multiply have two significant digits:

   A = 0.40 km · 0.28 km = 0.112 km² ≈ 0.11 km²

5. a. 3.0 in · 10,000 = 30,000 in = 30,000 in· (1 ft)/(12 in) · (1 mi)/(5,280 ft) = 0.473$\overline{48}$ mi ≈ 0.47 mi

   b. 3.0 in · 10,000 = 30,000 in = 30,000 in · (1 yd)/(36 in) = 833.$\overline{3}$ yd = ≈ 830 yd

6. Since 45.0 m and 21.2 have three significant digits and the scale ratio also has three, we will give the answers to three significant digits.

   45.0 m ÷ 500 = 0.09 m = 9.00 cm

   31.2 m ÷ 500 = 0.0624 m = 6.24 cm

   The dimensions on the map are 9.00 cm by 6.24 cm.

Review, p. 45

1. There is 4 m − (7 × 0.56 m) = 4 m − 3.92 m = 0.08 m left.

2.

a. 0.9 m = 90 cm 45 cm = 0.45 m 1.5 km = 1,500 m	b. 0.6 L = 600 ml 5,694 ml = 5.694 L 0.09 L = 90 ml	c. 2.2 kg = 2,200 g 390 g = 0.390 kg 0.02 kg = 20 g

3.

a. 6 ft 11 in. = 83 in. 3 lb 11 oz = 59 oz 3 C = 24 oz	b. 2 gal = 32 C 5 qt = 10 pt 54 oz = 6 C 6 oz	c. 78 oz = 4 lb 14 oz 39 in = 3 ft 3 in 102 in = 8 ft 6 in

4. The 40 yards of rope cost $15.99 / 40 yd = $0.40 per yard, and the 100 meters cost $40 ÷ 100 = $0.40 per meter. A meter is longer than a yard so the 100 meters is the better deal.

Review, cont.

5.

Prefix	Meaning	Units - length	Units - mass	Units - volume
centi-	hundredth = 0.01	centimeter (cm)	centigram (cg)	centiliter (cl)
deci-	tenth = 0.1	decimeter (dm)	decigram (dg)	deciliter (dl)
deca-	ten = 10	decameter (dam)	decagram (dag)	decaliter (dal)
hecto-	hundred = 100	hectometer (hm)	hectogram (hg)	hectoliter (hl)

6. a. 34 dl = 3.4 L
   b. 89 cg = 0.89 g
   c. 16 kl = 16,000 L

7.

a. 2.65 mi = 13,992 ft 10.9 mi = 19,184 yd	b. 3,800 ft = 0.72 mi 3,500 yd = 1.99 mi	c. 4.54 lb = 72.64 oz 10.2 ft = 122.4 in

8. a. The scale 1:45,000 means that 1 cm corresponds to 45,000 cm, which equals 450 m. So we can rewrite the scale as 1 cm : 450 m. Then, 850 m corresponds to 850/450 cm ≈ <u>1.9 cm</u>.

   Another way to solve this is to first divide 850 m ÷ 45,000 = 0.01$\overline{8}$ m, and then convert 0.01$\overline{8}$ m into centimeters: 0.01$\overline{8}$ m = 1.$\overline{8}$ cm ≈ 1.9 cm.

   b. 5.4 cm on the map corresponds to 5.4 cm · 45,000 = 243,000 cm = 2,430 m = <u>2.43 km</u> in reality.
   Another way to solve this is to use the converted scale of 1 cm : 450 m to get 5.4 cm · 450 m = 2,430 m = 2.43 km.

9. a. 2 digits   b. 3 digits   c. 4 digits   d. 1 digit   e. 4 digits   f. 5 digits
   g. 4 digits   h. 3 digits   i. 6 digits   j. 1 digit   k. 2 digits   l. 1 digit

10. a. 2.7 L   = <u>27 dl</u>   = <u>270 cl</u>   = <u>2700 ml</u>
    b. 5,600 m = <u>5.6 km</u>   = <u>56,000 dm</u> = <u>560,000 cm</u>
    c. 676 g   = <u>6,760 dg</u> = <u>67,600 cg</u> = <u>676,000 mg</u>

11.

a. 134 lb = 134 lb · $\dfrac{1 \text{ kg}}{2.2 \text{ lb}}$ = $\dfrac{134 \text{ kg}}{2.2}$ ≈ 60.9 kg
b. 156 cm = 156 cm · $\dfrac{1 \text{ in}}{2.54 \text{ cm}}$ · $\dfrac{1 \text{ ft}}{12 \text{ in}}$ = $\dfrac{156 \text{ ft}}{2.54 \cdot 12}$ ≈ 5.1 ft

12. The total capacity is 6 × 0.35 L + 2 × 2 L + 3 × 0.9 L = 2.1 L + 4 L + 2.7 L = <u>8.8 L</u>.

13. a. 56 oz = <u>3.5</u> lb   c. 2.7 gal = <u>10.8</u> qt   e. 0.48 mi = <u>2,534.4</u> ft
    b. 134 in = <u>11.17</u> ft   d. 0.391 lb = <u>6.26</u> oz   f. 2.45 ft = <u>2 ft 5.4 in</u>

Measurement Unit Conversions
Alignment to the Common Core Standards

The table below lists each lesson, and next to it the relevant Common Core Standard.

Lesson	page number	Standards
Decimals in Measuring Units and More	11	5.NBT.4 5.MD.1
Rounding and Estimating	15	5.NBT.4 5.MD.1
The Metric System	17	5.MD.1
Convert Metric Measuring Units	20	
Converting Between Customary Units of Measurement	23	5.NBT.4 5.MD.1
Convert Customary Measuring Units	27	
Convert Between Customary and Metric	31	
Using Ratios to Convert Measuring Units	33	6.RP.3
Maps	37	7.RP.3 7.NS.3
Significant Digits	43	7.RP.3 7.NS.3
Review	45	

Made in the USA
Lexington, KY
09 December 2019